Springer Theses

Recognizing Outstanding Ph.D. Research

Aims and Scope

The series "Springer Theses" brings together a selection of the very best Ph.D. theses from around the world and across the physical sciences. Nominated and endorsed by two recognized specialists, each published volume has been selected for its scientific excellence and the high impact of its contents for the pertinent field of research. For greater accessibility to non-specialists, the published versions include an extended introduction, as well as a foreword by the student's supervisor explaining the special relevance of the work for the field. As a whole, the series will provide a valuable resource both for newcomers to the research fields described, and for other scientists seeking detailed background information on special questions. Finally, it provides an accredited documentation of the valuable contributions made by today's younger generation of scientists.

Theses may be nominated for publication in this series by heads of department at internationally leading universities or institutes and should fulfill all of the following criteria

- They must be written in good English.
- The topic should fall within the confines of Chemistry, Physics, Earth Sciences, Engineering and related interdisciplinary fields such as Materials, Nanoscience, Chemical Engineering, Complex Systems and Biophysics.
- The work reported in the thesis must represent a significant scientific advance.
- If the thesis includes previously published material, permission to reproduce this must be gained from the respective copyright holder (a maximum 30% of the thesis should be a verbatim reproduction from the author's previous publications).
- They must have been examined and passed during the 12 months prior to nomination.
- Each thesis should include a foreword by the supervisor outlining the significance of its content.
- The theses should have a clearly defined structure including an introduction accessible to new PhD students and scientists not expert in the relevant field.

Indexed by zbMATH.

Atul Sharma

Twistor Sigma Models

Gravity, Amplitudes, and Flat Space Holography

Doctoral Thesis accepted by
University of Oxford, Oxford, UK

Springer

Author
Dr. Atul Sharma
Harvard University
Cambridge, MA, USA

Supervisor
Prof. Lionel Mason
The Mathematical Institute
University of Oxford
Oxford, UK

ISSN 2190-5053 ISSN 2190-5061 (electronic)
Springer Theses
ISBN 978-3-031-50750-2 ISBN 978-3-031-50751-9 (eBook)
https://doi.org/10.1007/978-3-031-50751-9

This Springer imprint is published by the registered company Springer Nature Switzerland AG
The registered company address is: Gewerbestrasse 11, 6330 Cham, Switzerland

Paper in this product is recyclable.

To my parents

Supervisor's Foreword

This thesis is a unique work at the intersection of scattering amplitudes, holography and twistor theory that promises to have repercussions across all these fields. Scattering amplitudes generate the probabilities of the interaction processes of fundamental particles that are predicted by quantum field theories. They are central both to the understanding of the physics seen in colliders and their underlying mathematical structures.

In recent years, perturbative approaches to scattering amplitudes in gravity have received new impetus both from observation and theory. On the one hand, gravitational wave detectors need new techniques for the computation of gravitational wave signatures arising from black hole and neutron star mergers and on the other, recent advances in scattering amplitudes and holography suggest new and more profound formulations of Einstein gravity. This thesis focuses on twistor approaches to Einstein gravity in four dimensions with a focus on the construction of scattering amplitudes and their connections with holography. It represents a breakthrough in the area via the introduction of a novel sigma model in twistor space that directly leads to simple formulae for scattering amplitudes. These simplify previous approaches considerably, having a more direct connection with the space-time action, and requiring only classical, rather than quantum calculations in the first instance. It leads to a number of applications.

The first-ever top-down example of holography in an asymptotically flat space-time was finally discovered last year by Costello, Paquette and the author of this thesis. The work presented in this thesis was key in bringing about the paradigm shift that precipitated this discovery. As a result, this thesis forms essential reading for any graduate students or young researchers interested in state-of-the-art advances in flat holography and scattering amplitudes.

Twistor sigma models are 2-dimensional chiral conformal field theories living on 2-spheres determining their embedding into a twistor space, a three complex-dimensional space. They provide a 2-dimensional theory that is dual to the 4-dimensional S-matrix of general relativity. They compute graviton scattering amplitudes in flat space. But they also work in exciting new settings like self-dual curved space-times where all the tricks in the arsenal of a normal amplitudologist inevitably fail.

The secret to their success are precisely the ongoing leaps in flat space holography. Twistor sigma models underpin the recently uncovered celestial $w_{1+\infty}$ symmetry of quantum gravity. But they go further and turn it into a dynamical computational tool, leveraging it for obtaining graviton amplitudes as well as classical solutions of self-dual gravity.

Since this thesis came out, many further advances have shown that twistor sigma models form a crucial building block of top-down celestial holography. They arise as worldvolume theories of D1 branes in the topological B model string theory on twistor space, and thereby enter the celestial CFT duals of self-dual gauge and gravitational theories in asymptotically flat space-times such as Burns space.

As such, twistor sigma models promise to remain an irreplaceable tool for exploring the physics and mathematics of quantum gravity for many years to come. This thesis is mostly rewritten and greatly expands upon the sections of the papers on which it is based that it discusses with improved pedagogy.

In summary, this thesis is a significant and substantial contribution within and across fields, and it will be a valuable resource for both physicists and mathematicians investigating geometrical structures emerging from the physics of amplitudes, flat space holography and twistor theory for some time to come.

Oxford, UK
October 2023

Prof. Lionel Mason

Abstract

This thesis presents an overview of recent developments in the applications of twistor theory to the study of the gravitational S-matrix in flat as well as curved space-times. We begin by introducing the novel geometric tool of *twistor sigma models*. These are 2-dimensional chiral sigma models governing holomorphic maps from the Riemann sphere into twistor spaces of self-dual vacuum space-times. What follows is a concise list of their main highlights, in the order that the reader will encounter them:

- Solutions to the equations of motion of our sigma models provide the incidence relations of the twistor correspondence.
- On-shell actions of our models compute Kähler and Plebanski potentials for hyperkähler metrics on the associated space-times.
- Connected, tree-level correlators of our models give rise to tree-level, maximally helicity violating (MHV) graviton amplitudes of general relativity in flat space.
- The chiral algebra of operators in our models enjoys an action of the loop algebra of the wedge subalgebra of $w_{1+\infty}$. This is associated with the soft sector of celestial holography.
- By coupling them to background twistor spaces of self-dual, vacuum space-times, our models can be used to derive tree-level MHV graviton amplitudes in self-dual, radiative space-times.

The self-dual space-times mentioned in the last point provide an ideal laboratory for studying amplitudes in curved backgrounds. The corresponding formulae for MHV graviton amplitudes are built out of Hodges' determinants familiar from flat space, but also exhibit exciting new structures like gravitational wave tails.

Acknowledgements

I am grateful to my supervisor, Lionel Mason, for his invaluable lessons on geometry, physics and cafés, and to my mentors and collaborators Tim Adamo, Roland Bittleston, Eduardo Casali and David Skinner for their endless support. I would also like to thank all my wonderful friends from Oxford and Cambridge for numerous insightful discussions and cherished days in the sun.

My research at Oxford has been supported by a Mathematical Institute Studentship, and near its end by Francis Brown's ERC grant GALOP ID: 724638.

Publications related to this thesis

1. T. Adamo, L. Mason, and A. Sharma, *Twistor sigma models for quaternionic geometry and graviton scattering*, arXiv:2103.16984
2. A. Sharma, *Twistor action for general relativity*, arXiv:2104.07031
3. R. Bittleston, A. Sharma, and D. Skinner, *Quantizing the non-linear graviton*, arXiv:2208.12701
4. T. Adamo, L. Mason, and A. Sharma, *Celestial $w_{1+\infty}$ Symmetries from Twistor Space*, *SIGMA* **18** (2022) 016, [arXiv:2110.06066]
5. T. Adamo, W. Bu, E. Casali, and A. Sharma, *Celestial operator products from the worldsheet*, arXiv:2111.02279
6. T. Adamo, L. Mason, and A. Sharma, *MHV scattering of gluons and gravitons in chiral strong fields*, *Phys. Rev. Lett.* **125** (2020), no. 4 041602, [arXiv:2003.13501]
7. T. Adamo, L. Mason, and A. Sharma, *Graviton scattering in self-dual radiative space-times*, arXiv:2203.02238

Contents

Chapter 1
Introduction

1.1 Dualities in the Real World

Dualities abound at the frontiers of physics. They help us relate aspects of one physical system to another, often leading to giant leaps in conceptual understanding as well as calculational prowess. The most famous example of these are the holographic dualities in string theory that relate gravitational theories on one space to non-gravitational theories on another space of lower dimension [1]. Weak coupling observables like scattering amplitudes on the gravitational side map to strong coupling correlation functions in the holographic dual. Most examples of this nature belong to the class of AdS/CFT dualities, where the bulk gravitational theory naturally lives in asymptotically anti-de Sitter space-times. They provide a concrete definition of quantum gravity in AdS in terms of non-gravitational CFTs on its boundary.

To date, similar searches for holographic duals of quantum gravity in asymptotically flat space-times have met with little success. Instead, the second half of the twentieth century saw the rise of a different class of dualities that relate weak coupling physics on 4-dimensional flat space to weak coupling physics on a complex 3-fold known as *twistor space* [2]. They reorganize the physics of massless particles into the extremely efficient machinery of complex structure deformation theory. Over the years, twistor theory has grown into a highly successful paradigm for describing the classical and perturbative regimes of gauge theory and gravity [3]. Ever since Witten's discovery of the twistor string [4], it has revolutionized the study of scattering amplitudes. This has given rise to a multitude of remarkable all-multiplicity formulae for amplitudes of gluons, gravitons, and other massless particles (see for example [5] for a modern review).

Meanwhile, the hunt for flat space holography has been picking up again. Recent work on asymptotic symmetries, dubbed *celestial holography* [6, 7], has taught us that holographic duals of gauge theory and quantum gravity in 4d Minkowski space might find a natural home on the 2d celestial sphere. A promising direction toward discovering such a 2d dual arises in the study of holography applied to twistor strings

A. Sharma, *Twistor Sigma Models*, Springer Theses,
https://doi.org/10.1007/978-3-031-50751-9_1

[8]. At their core, twistor strings are examples of topological strings on complex 3-folds, and holographic duals of the latter indeed tend to be the worldvolume theories of D1-branes wrapped on Riemann spheres [9–12]. This procedure has recently led to the first successful discovery of a celestial dual pair in [13, 14], although this has not yet captured realistic models of gravity.

Motivated by the search for such a dual 2d description of flat space gravity, in this thesis we will pursue a general class of 2d chiral (defect) conformal field theories (CFTs) referred to as *twistor sigma models*. Their actions roughly take the form

$$S = \frac{1}{2\pi \mathrm{i}} \int_{\mathbb{CP}^1} \frac{\mathrm{d}z\,\mathrm{d}\bar{z}}{z^2} \left(m^{\dot{2}} \bar{\partial}_{\bar{z}} m^{\dot{1}} + h(m) \right) \tag{1.1}$$

where $m^{\dot{\alpha}} = (m^{\dot{1}}, m^{\dot{2}})$ are a pair symplectic bosons, and h encodes coupling to positive helicity gravitons. They govern deformations of holomorphically embedded Riemann spheres in twistor space under deformations of the ambient complex structure. The measure has second order poles at $z = 0, \infty$ which act as defects in twistor space. Using these models, we will find dual incarnations of graviton scattering amplitudes as 2d correlators, leading to a new derivation of maximally helicity violating (MHV) graviton tree-amplitudes. They also yield new insights into flat space holography, unraveling an infinite dimensional holographic symmetry algebra that potentially underlies the remarkable simplicity of MHV amplitudes.

Our twistor sigma models originally emerged in the study of graviton scattering in curved space-times. In flat space, twistor string theory has given rise to a multitude of all-multiplicity worldsheet formulae for amplitudes of gluons [15] and gravitons [16, 17]. In an MHV configuration involving negative helicity gravitons $1, 2$ and positive helicity gravitons $3, \ldots, n$, the tree-level graviton amplitude is captured by an extremely elegant formula due to Andrew Hodges [18],

$$\frac{\langle 1\,2\rangle^6}{\langle 1\,r\rangle^2\,\langle 2\,r\rangle^2}\,|\mathbb{H}^r_r|\,. \tag{1.2}$$

Here, $|\cdot|$ denotes the determinant, and $\mathbb{H}^r_r$ is the $(n-3)\times(n-3)$ matrix obtained by removing the row and column corresponding to the rth graviton from "Hodges' matrix" $\mathbb{H}$:

$$\mathbb{H}_{ij} = \frac{[i\,j]}{\langle i\,j\rangle}\,, \quad i \neq j\,; \qquad \mathbb{H}_{ii} = -\sum_{j\neq i} \frac{[i\,j]}{\langle i\,j\rangle}\,\frac{\langle 1\,j\rangle\,\langle 2\,j\rangle}{\langle 1\,i\rangle\,\langle 2\,i\rangle}\,, \tag{1.3}$$

where $i, j = 3, 4, \ldots, n$. The objects $[i\,j]$ and $\langle i\,j\rangle$ are certain Lorentz invariant contractions built out of the graviton momenta that will be introduced in due course. Hodges' formula seems ever the more ingenious when one realizes that it sums millions of Feynman diagrams generated by the perturbative expansion of general relativity (GR) even for the scattering of relatively few gravitons.

Nonetheless, almost all of the progress in finding such formulae has been made in flat space. In contrast, recent decades have seen the rise of perturbative calculations in quantum field theories on numerous curved backgrounds. These range from those of purely theoretical interest like anti-de-Sitter spaces, to those spanning real-world applications like in cosmology, black holes, and gravitational wave physics. Amplitudes in curved backgrounds are also interesting from a mathematical standpoint, as they display many novel functional and geometrical features that are absent in trivial backgrounds.

This brings us to one of the overarching questions of this thesis:

Is the simplicity of the S-matrix in flat space
an intrinsic feature of perturbative quantum gravity?

There is a very good argument for why the answer to this question could be "no". Flat backgrounds are maximally symmetric, so it is possible that scattering in flat space is maximally nice because it is maximally constrained by global and gauge symmetries, along with locality and unitarity. In curved backgrounds, one generically has little to no global symmetries. And the effects of locality and unitarity are poorly understood since amplitudes in curved backgrounds are no longer rational functions of momenta even at tree-level. This means that many modern "on-shell methods" of computing amplitudes [19, 20] like BCFW recursion, generalized unitarity, double copy, etc. are very hard to extend to curved backgrounds as they stand.

In this work, we will analyze these questions by computing amplitudes in self-dual (SD) curved space-times. Self-duality provides a perfect laboratory to study how turning on curvature affects various observables, since we can use the integrability of self-dual gravity to our advantage. In such backgrounds, twistor strings can in principle overcome the aforementioned hurdles. But the resulting worldsheet formulae for the amplitudes would remain conjectural, as they cannot be verified using BCFW recursion as is usually done in flat space [21]. At this stage, our twistor sigma models come to the rescue. Starting directly from the Einstein-Hilbert action, we will prove from first principles that semi-classical correlators of our models compute tree-level graviton MHV amplitudes in both flat and self-dual curved backgrounds. This will bypass the need for a separate proof via on-shell recursion.

To be completely explicit, we will study graviton amplitudes on self-dual *radiative* space-times. Appropriate notions of momentum eigenstates, null infinity and the S-matrix continue to persist in such backgrounds. The resulting formulae for MHV amplitudes show many new structures like gravitational wave tails, but also continue to display remarkable simplicity much as in the flat space case. In fact, we can extend our sigma models to even conjecture formulae for the non-MHV tree-amplitudes on such backgrounds, though a direct proof of these as in the MHV case remains out of reach. As this is a subject of active investigation, we will not discuss it here; see [22] for more details.

1.2 Thesis Summary

In the rest of this introduction, we briefly summarize the main results that the reader may look forward to.

Chapter 2 provides a comprehensive review of the twistor theory for flat space. Over the last few decades, twistor space has become an essential backdrop of most modern studies of scattering amplitudes. Most other reviews of the subject (as applied to physics) focus on twistor theory in Lorentzian or split, i.e., (2, 2) signature. We choose to review a complementary approach that has played an equally important role: twistor theory in Euclidean signature. Traditionally, this is the facet of twistor theory that found applications to differential geometry. But it also feeds back into studies of quantum field theory through the construction of instantons as well as scattering amplitudes.

Starting with the twistor theory for complexified space-time $\mathbb{C}^4$, we quickly specialize to the case of Euclidean $\mathbb{R}^4$. The corresponding twistor space is an open subset of complex projective 3-space that we denote $\mathbb{PT}$. We review the Penrose transform that maps solutions of massless free field equations on $\mathbb{R}^4$ to certain cohomology classes on $\mathbb{PT}$ known as twistor representatives. These provide the external wave-functions that will enter the computation of space-time amplitudes via theories on twistor space. Other helpful reviews of Euclidean twistor theory include [23–26].

Chapter 3 starts by reviewing the tools of curved twistor theory for four dimensional Riemannian manifolds with self-dual vacuum (i.e., Ricci flat) metrics. This goes under the umbrella of Penrose's non-linear graviton construction [27]. A self-dual vacuum space-time M is necessarily hyperkähler. That is, it admits a triplet of metric compatible complex structures I, J, K that satisfy the quarternion algebra $I^2 = J^2 = K^2 = IJK = -1$. Such space-times then admit a whole 2-sphere's worth of complex structures $aI + bJ + cK$, one for each point $(a, b, c) \in \mathbb{R}^3$ on the unit 2-sphere $a^2 + b^2 + c^2 = 1$. Twistor space, denoted $\mathbb{P}\mathcal{J}$, may be defined as a bundle of 2-spheres over M whose global sections are these complex structures (see [28] for a standard review), though we provide more practical definitions in the main text.

Picking a choice of complex structure, we can write the metric on M in terms of second derivatives of a Kähler potential Ω. This Kähler potential is the protagonist of our story. Ricci flatness of the associated Kähler metric translates into an extremely elegant PDE (partial differential equation) on Ω known as Plebanski's (first) heavenly equation [29]. In complex coordinates, the Ricci tensor of Kähler metrics can be expressed in terms of second derivatives of a logarithm of the metric determinant. So in its simplest avatar, the heavenly equation states that the Kähler metric has unit determinant. This turns into a second order PDE for Ω.

Much of the rest of Chap. 3 is concerned with developing a concrete algorithm for solving this PDE. We accomplish this by introducing the twistor sigma models in Sect. 3.3 that are the namesake of this work. As mentioned before, these are chiral sigma models whose target space is twistor space. Our main result, Theorem 3.2, states that one can use them to construct a "general" solution of the heavenly equation.[1] It starts with arbitrary gravitational twistor representatives h, to be thought of as "initial data" for the PDE, and produces Kähler potentials through the on-shell action of our sigma models. In fact, it may be thought of as a generalization of the Legendre transform construction of Hitchin et al. [30]. Unlike the Legendre transform, the twistor sigma model does not require any space-time isometries for it to work.

Chapter 4 deals with generalizations of our twistor sigma models to higher dimensions as well as to quaternion-Kähler geometry. This chapter may be safely skipped by the reader interested only in the physics of graviton scattering amplitudes. It is more oriented toward applications to algebraic and differential geometry. It starts by giving an ab initio construction of our sigma models from the symplectic structure on twistor space. The data of a choice of complex structure on space-time enters through a pair of source terms, and the on-shell action again computes the space-time Kähler scalar in the chosen complex structure.

In the sections that follow, we describe twistor sigma models that compute scalar potentials for hyperkähler metrics on $4k$-dimensional manifolds, $k \in \mathbb{Z}_{\geq 1}$. We consider two types of potentials: the Kähler scalar and a second type of potential known as the Plebanski scalar [29]. They each satisfy certain PDEs that encode Ricci-flatness, and we describe sigma models into $(2k + 1)$-complex-dimensional twistor spaces whose on-shell actions solve these PDEs. We also describe quaternion-Kähler metrics, which are a higher-dimensional generalization of self-dual Einstein metrics [31]. These are described in terms of a scalar potential known as the Przanowski scalar [32]. In fact, this lifts to a Kähler scalar known as the Swann scalar that induces a hyperkähler structure on non-projective twistor space [33]. The Swann scalar is again shown to be computed by a twistor sigma model with a non-zero cosmological constant Λ turned on.

In Chap. 5, we start building toward our main application of this formalism: the computation of scattering amplitudes in general relativity. In the past, most work on graviton scattering has concerned the application of on-shell recursion relations or double copy techniques. An introductory textbook treatment of these approaches is discussed in [19]. We provide a complementary approach by instead focusing directly on space-time perturbation theory starting from the action of GR.

We start by reviewing Plebanski's chiral action for GR, which happens to be equivalent to the Einstein-Hilbert action up to a topological term. In [34], this allowed Mason and Skinner to derive a generating functional for tree-level graviton MHV amplitudes. They found an uplift of this generating functional to twistor space, which turned out to be an efficient first-principles tool for generating new amplitude formulae. These formulae matched the existing formulae for MHV amplitudes due to Berends, Giele and Kuijf (BGK) [35], but were soon thereafter superseded in simplicity by Hodges' formulae [18]. The latter also led to the development of worldsheet formulae for graviton tree amplitudes in arbitrary helicity sectors [16], but the dynamical origin of Hodges' discovery remained a mystery.

[1] We work perturbatively but fully non-linearly around a flat background.

In what follows, we will resolve this mystery by finding a new generating functional that directly reproduces Hodges' instead of the Mason-Skinner or BGK formulae. To accomplish this, we recast the Mason-Skinner generating functional as a Fourier transform of a Kähler scalar Ω,

$$\int \mathrm{d}^4x \; \Omega(x)\, \mathrm{e}^{\mathrm{i}(k_1+k_2)\cdot x}\,. \tag{1.4}$$

The idea is to express the scattering in a geometrical guise. One thinks of the + helicity gravitons as forming a classical SD Ricci flat background in which the two − helicity gravitons 1, 2 scatter. The data of the + helicity gravitons is contained in the Kähler scalar Ω, while that of the − helicity ones is contained in their null momenta k_1, k_2 (as well as in the choice of complex structure).

The main tool that allows us to actually compute Ω, and thereby its Fourier transform, is the twistor sigma model for Ω. The sigma model computes Ω via its on-shell action, with the twistor representative h taken to be a linear superposition of representatives for the individual + helicity states. We recast this on-shell action in terms of connected, tree-level correlators of certain vertex operators for the sigma models. These correlators can be explicitly obtained using 2d CFT techniques and lead to a first-principles proof of Hodges' formula in flat space.

The connections of our sigma models with celestial holography are explored in Chap. 6. As mentioned briefly above, the goal of celestial holography is to find a 2d hologram for quantum gravity in asymptotically flat space-times. Such a hologram is often referred to as a celestial CFT (CCFT for short). Since our twistor sigma models indeed provide a 2d CFT that computes 4d MHV amplitudes, we argue that they capture a subsector of such a putative holographic dual.

The primary tools used to study celestial CFTs are celestial operator product expansions (OPEs) [36]. Every graviton state propagating in the bulk is posited to be dual to an operator in the CCFT. Celestial OPEs are the hypothetical OPEs that one gives these operators so as to reproduce collinear limits of bulk scattering amplitudes via OPE expansions of CCFT correlation functions. In particular, soft gravitons are gravitons of zero energy, and their dual operators generate chiral symmetry algebras of the CCFT.

In [37], Guevara et al. discovered an infinite dimensional chiral algebra governing the soft graviton OPEs. It was later recognized by Strominger [38] to be the loop algebra of the algebra of canonical transformations of a 2-plane. This is often referred to as the celestial $w_{1+\infty}$ algebra in the literature. It happens to precisely match the global symmetry algebra of self-dual GR [8, 39], a fact that only becomes manifest through the lens of Penrose's non-linear graviton construction! Because of this fact, it is expected to form a symmetry algebra of the celestial CFT dual to self-dual gravity. It also acts on the celestial OPE in the non-self-dual sector, but is possibly broken by the vacuum of full GR.

The semiclassical OPEs of our twistor sigma models provide a free field realization of such chiral algebras. Soft graviton vertex operators are shown to give rise to symmetry currents generating the loop algebra of the wedge subalgebra of $w_{1+\infty}$.

The twistor space of $\mathbb{R}^4$ can be viewed as a holomorphic $\mathbb{C}^2$ bundle over a Riemann sphere: $\mathbb{C}^2 \hookrightarrow \mathbb{PT} \to \mathbb{CP}^1$. In this picture, the soft graviton vertex operators are in one-to-one correspondence with a basis of hamiltonians generating holomorphic symplectomorphisms of the $\mathbb{C}^2$ fibers. And the 2d CFT coordinate $z \in \mathbb{CP}^1$ arises from the dependence of these hamiltonians on the base of this fibration. As a result, one can concretely interpret the CCFT as being a 2d chiral system placed along this $\mathbb{CP}^1$.

This of course leaves us with many open questions. The OPEs of our twistor sigma models start differing from the flat space celestial OPEs once multiple contractions (loop corrections in the 2d theory) are taken into account. Are these supposed to capture loop corrections in the bulk, or perhaps non-commutativity [40, 41]? Are there twistor sigma models that compute amplitudes in other helicity sectors? Can we find a dual for full GR by expanding around the dual for SDGR? A preliminary discussion of such questions may be found in the papers [39, 42], which may form useful jumping points of exploration for the intrigued reader.

Following this, we come to perhaps the most novel application of our formalism: the study of amplitudes in curved backgrounds. Twistor sigma models work best on self-dual space-times, so it is these backgrounds that we restrict attention to. For computational ease, we also simplify life by working in the category of radiative space-times. In Chap. 7, we review the notion of a self-dual radiative space-time and set up the associated twistor spaces.

A radiative space-time is a Lorentzian (or complexified) space-time whose metric is entirely determined by initial data at null infinity. That is, such metrics are source-free solutions of the Einstein equations. These space-times are useful because they allow for decent notions of null infinities and scattering amplitudes to persist beyond the flat space case. We complexify such space-times and further specialize to SD radiative space-times.[2]

The twistor spaces of such backgrounds can be explicitly constructed. And they happen to be particularly amenable to supporting twistor sigma models with a tractable perturbation theory. The Penrose transform also allows us to find exact solutions of the wave equations on SD radiative space-times, which once again feed into the construction of vertex operators of the sigma models.

Chapter 8 describes how to set up our sigma models on such background twistor spaces. As of this writing, this has not been carried out in the greatest generality for arbitrary SD vacuum space-times, but it is relatively straightforward to do on radiative ones. The main new ingredient one finds is the presence of infinitely many non-linear interaction vertices that encode scattering of gravitons off the background itself. The sigma model action schematically takes the form

$$S = \frac{1}{2\pi \mathrm{i}} \int_{\mathbb{CP}^1} \frac{\mathrm{d}z\,\mathrm{d}\bar{z}}{z^2} \left(m^{\dot{2}} \bar{\partial}_{\bar{z}} m^{\dot{1}} + h(m) + \sum_{p=2}^{\infty} \mathsf{h}_{\dot{\alpha}_1 \cdots \dot{\alpha}_p} m^{\dot{\alpha}_1} \cdots m^{\dot{\alpha}_p} \right) \tag{1.5}$$

---

[2] These space-times may not generically have a Euclidean real slice, but twistor theory works regardless.

where $\mathsf{h}_{\dot\alpha_1\cdots\dot\alpha_p}$ are certain derivatives of the twistor representative h of the background that play the role of coupling constants.

The $p = 2$ term in the sum deforms the kinetic term, whereas the cubic and higher order self-couplings of $m^{\dot\alpha}$ give rise to the so-called *tail terms* in graviton amplitudes. These are physically interpreted as gravitons being absorbed into the background and then being subsequently reemitted as gravitational wave tails. When computing the amplitudes from correlators of this 2d theory, one needs to account for insertions of all of these new vertices. We show that this can be done using simple generalizations of the combinatorial techniques used to evaluate the flat space graviton amplitude.

Our quest thus ends with the discovery of the all-multiplicity formula (8.64) for MHV graviton tree-amplitudes on all SD radiative backgrounds. This formula is neatly packaged in terms of the Bondi news tensor $N(u, z, \bar{z})$ of the background, with high-order tail effects coming with higher and higher derivatives of the news. It reduces to precisely Hodges' formula in the flat space limit as the background is turned off. We also mention other consistency checks like matching with Feynman diagrammatics, as well as a linearization of our formula in the background which leads to the notion of a *perturbative limit*. In the latter case, the background can be Fourier transformed, and each Fourier mode can itself be treated as an extra positive helicity graviton participating in the scattering process. We explicitly show that the result of such linearization is exactly the flat space MHV amplitude containing one extra graviton representing the background.

1.3 Further Reading

Since the original writing of this thesis, a host of exciting new developments have taken place in the field of twistor sigma models and celestial holography that are worth mentioning.

The idea of twistor sigma models is slowly being understood to be a special case of the more general concept of *celestial chiral algebras*. The latter were introduced in beautiful works of Strominger [38], followed by Costello and Paquette [8], as a giant leap toward the bottom-up understanding of celestial holography in the self-dual sectors of gauge theory and gravity. Subsequently, these lines of thought also converged upon the first-ever top-down celestial duality in the author's work [13, 14]. This duality relates certain gauge-fixed models of self-dual Yang-Mills and self-dual conformal gravity on an asymptotically Euclidean, scalar-flat Kähler 4-manifold called Burns space, to an explicit chiral algebra (chiral 2d CFT) living on a stack of D1 branes wrapping a 2-sphere in twistor space. Our twistor sigma models (for flat space) form a subsector of this chiral algebra governing the center-of-mass motion of the D1 branes [43, 44].

The celestial chiral algebra for self-dual GR has been analyzed at both the classical and quantum levels in [45]. In the semiclassical limit, our twistor sigma models provide a free field realization of this algebra, but the question of their extension to quantum self-dual gravity is under active investigation. In particular, this requires a careful study of anomaly cancellation in certain quantum field theories on twistor space [43, 46], which translates to the associativity of their celestial duals [45, 47]. The celestial chiral algebras have also been studied on interesting non-radiative self-dual space-times like Eguchi-Hanson space [41], and have been realized to be avatars of the kinematic algebra of doubly copy fame in self-dual sectors [48–51]. It would be interesting to see how these various lines of work converge upon an understanding of quantum gravity in the real world as time progresses.

References

1. Aharony O, Gubser SS, Maldacena JM, Ooguri H, Oz Y (2000) Large N field theories, string theory and gravity. Phys Rep 323:183–386 arXiv:hep-th/9905111
2. Penrose R, MacCallum MAH (1972) Twistor theory: an approach to the quantization of fields and space-time. Phys Rep 6:241–316
3. Atiyah M, Dunajski M, Mason L (2017) Twistor theory at fifty: from contour integrals to twistor strings. Proc R Soc Lond A 473(2206):20170530. arXiv:1704.07464
4. Witten E (2004) Perturbative gauge theory as a string theory in twistor space. Commun Math Phys 252:189–258 arXiv:hep-th/0312171
5. Geyer Y, Mason L The SAGEX review on scattering amplitudes, Chapter 6: ambitwistor strings and amplitudes from the worldsheet. arXiv:2203.13017
6. Strominger A Lectures on the infrared structure of gravity and gauge theory. arXiv:1703.05448
7. Pasterski S, Pate M, Raclariu A-M (2021) Celestial holography. In: 2022 Snowmass summer study, vol 11. arXiv:2111.11392
8. Costello K, Paquette NM Celestial holography meets twisted holography: 4d amplitudes from chiral correlators. arXiv:2201.02595
9. Dijkgraaf R, Vafa C (2002) Matrix models, topological strings, and supersymmetric gauge theories. Nucl Phys B 644:3–20. arXiv:hep-th/0206255
10. Bonetti F, Rastelli L (2018) Supersymmetric localization in AdS_5 and the protected chiral algebra. JHEP 2018(8):98. arXiv:1612.06514
11. Costello K, Gaiotto D Twisted holography. arXiv:1812.09257
12. Ishtiaque N, Moosavian SF, Zhou Y (2020) Topological holography: the example of the D2-D4 brane system. SciPost Phys 9(2):017. arXiv:1809.00372
13. Costello K, Paquette NM, Sharma A (2023) Top-down holography in an asymptotically flat spacetime. Phys Rev Lett 130(6):061602. arXiv:2208.14233
14. Costello K, Paquette NM, Sharma A Burns space and holography. arXiv:2306.00940
15. Roiban R, Spradlin M, Volovich A (2004) On the tree level S matrix of Yang-Mills theory. Phys Rev D 70:026009. arXiv:hep-th/0403190
16. Cachazo F, Skinner D (2013) Gravity from rational curves in twistor space. Phys Rev Lett 110(16):161301. arXiv:1207.0741
17. Skinner D (2020) Twistor strings for N = 8 supergravity. JHEP 04:047. arXiv:1301.0868
18. Hodges A A simple formula for gravitational MHV amplitudes. arXiv:1204.1930
19. Elvang H, Huang Y-T Scattering amplitudes. arXiv:1308.1697
20. Travaglini G et al The SAGEX review on scattering amplitudes. arXiv:2203.13011
21. Cachazo F, Mason L, Skinner D (2014) Gravity in twistor space and its Grassmannian formulation. SIGMA 10:051. arXiv:1207.4712
22. Adamo T, Mason L, Sharma A, Graviton scattering in self-dual radiative space-times. arXiv:2203.02238
23. Woodhouse NMJ (1985) Real methods in twistor theory. Class Quantum Gravity 2:257–291
24. Jiang W (2008) Aspects of Yang-Mills theory in twistor space. PhD thesis, University of Oxford. arXiv:0809.0328

25. Adamo T (2018) Lectures on twistor theory. PoS Modave2017:003. arXiv:1712.02196
26. Adamo T (2013) Twistor actions for gauge theory and gravity. PhD thesis, University of Oxford. arXiv:1308.2820
27. Penrose R (1976) The nonlinear graviton. Gen Relativ Gravit 7:171–176
28. Hitchin N (1992) Hyperkähler manifolds, Sem. Bourbaki, Asterisque 206:137–166
29. Plebanski JF (1975) Some solutions of complex Einstein equations. J Math Phys 16:2395–2402
30. Hitchin NJ, Karlhede A, Lindstrom U, Rocek M (1987) Hyperkahler metrics and supersymmetry. Commun Math Phys 108:535
31. Salamon S (1982) Quaternionic Kähler manifolds. Inven Math 67:143–171
32. Przanowski M (1983) Locally Hermite Einstein, selfdual gravitational instantons. Acta Phys Pol B 14:625–627
33. Swann A (1991) Hyper-Kähler and quaternionic Kähler geometry. Math Ann 289(3):421–450
34. Mason LJ, Skinner D (2010) Gravity, twistors and the MHV formalism. Commun Math Phys 294:827–862. arXiv:0808.3907
35. Berends FA, Giele WT, Kuijf H (1988) On relations between multi-gluon and multi-graviton scattering. Phys Lett B 211:91–94
36. Pate M, Raclariu A-M, Strominger A, Yuan EY Celestial operator products of gluons and gravitons. arXiv:1910.07424
37. Guevara A, Himwich E, Pate M, Strominger A Holographic symmetry algebras for gauge theory and gravity. arXiv:2103.03961
38. Strominger A w(1+infinity) and the celestial sphere. arXiv:2105.14346
39. Adamo T, Mason L, Sharma A (2022) Celestial $w_{1+\infty}$ symmetries from twistor space. SIGMA 18:016. arXiv:2110.06066
40. Bu W, Heuveline S, Skinner D (2022) Moyal deformations, $\mathrm{W}_{1+\infty}$ and celestial holography. JHEP 12:011. arXiv:2208.13750
41. Bittleston R, Heuveline S, Skinner D The celestial chiral algebra of self-dual gravity on Eguchi-Hanson space. arXiv:2305.09451
42. Adamo T, Mason L, Sharma A Twistor sigma models for quaternionic geometry and graviton scattering. arXiv:2103.16984
43. Costello KJ Quantizing local holomorphic field theories on twistor space. arXiv:2111.08879
44. Costello K, Li S (2020) Anomaly cancellation in the topological string. Adv Theor Math Phys 24(7):1723–1771. arXiv:1905.09269
45. Bittleston R (2023) On the associativity of 1-loop corrections to the celestial operator product in gravity. JHEP 01:018. arXiv:2211.06417
46. Bittleston R, Sharma A, Skinner D Quantizing the non-linear graviton. arXiv:2208.12701
47. Costello K, Paquette NM (2022) Associativity of one-loop corrections to the celestial operator product expansion. Phys Rev Lett 129(23):231604. arXiv:2204.05301
48. Monteiro R (2023) Celestial chiral algebras, colour-kinematics duality and integrability. JHEP 01:092. arXiv:2208.11179
49. Monteiro R, Stark-Muchão R, Wikeley S, Anomaly and double copy in quantum self-dual Yang-Mills and gravity. arXiv:2211.12407
50. Monteiro R (2023) From Moyal deformations to chiral higher-spin theories and to celestial algebras. JHEP 03:062. arXiv:2212.11266
51. Brown GR, Gowdy J, Spence B Self-dual fields on self-dual backgrounds and the double copy. arXiv:2307.11063
52. He T, Lysov V, Mitra P, Strominger A (2015) BMS supertranslations and Weinberg's soft graviton theorem. JHEP 05:151. arXiv:1401.7026

Chapter 2
Twistors for Flat Space

We begin with a short review of the construction of twistor space for complexified four-dimensional Minkowski space. Following this, we specialize to the twistor geometry of Euclidean signature flat space. We will utilize a judicious combination of both viewpoints in this thesis.

2.1 Twistor Correspondence

Let $x^a = (x^0, x^1, x^2, x^3)$ be complex coordinates on $\mathbb{C}^4$. We work with a "Minkowski metric" on $\mathbb{C}^4$, by which we mean the holomorphic quadratic differential

$$\mathrm{d}s^2 = (\mathrm{d}x^0)^2 - (\mathrm{d}x^1)^2 - (\mathrm{d}x^2)^2 - (\mathrm{d}x^3)^2\,. \tag{2.1}$$

Although this is not an actual metric on $\mathbb{C}^4$, it generically gives rise to flat metrics of various signatures on real slices $\mathbb{R}^4 \subset \mathbb{C}^4$. Directly working on $\mathbb{C}^4$ allows us to treat all signatures simultaneously, while any specific signature can be picked out by imposing certain "reality conditions" that we discuss below.

We introduce spinors via the decomposition $\mathrm{SO}(4,\mathbb{C}) \simeq (\mathrm{SL}(2,\mathbb{C}) \times \mathrm{SL}(2,\mathbb{C}))/\mathbb{Z}_2$ of the structure group of the tangent bundle of $\mathbb{C}^4$. Concretely, we construct spinorial coordinates on $\mathbb{C}^4$ as

$$x^{\alpha\dot\alpha} = \frac{1}{\sqrt{2}}\begin{pmatrix} x^0 + x^3 & x^1 - \mathrm{i}\,x^2 \\ x^1 + \mathrm{i}x^2 & x^0 - x^3 \end{pmatrix}, \qquad \alpha = 1, 2,\ \dot\alpha = \dot{1}, \dot{2}. \tag{2.2}$$

Spinor components of any tensor on $\mathbb{C}^4$ are then defined as its components in the bases $\{\partial_{\alpha\dot\alpha} \equiv \partial/\partial x^{\alpha\dot\alpha}\}$, $\{\mathrm{d}x^{\alpha\dot\alpha}\}$ of the tangent and cotangent bundles [1]. The 2-spinor indices α, $\dot\alpha$ transform in the fundamental representations of the two $\mathrm{SL}(2,\mathbb{C})$

A. Sharma, *Twistor Sigma Models*, Springer Theses,
https://doi.org/10.1007/978-3-031-50751-9_2

factors. They can be raised or lowered using Levi-Civita symbols $\epsilon_{\alpha\beta}$, $\epsilon_{\dot\alpha\dot\beta}$. We will work with the conventions

$$\epsilon_{\alpha\beta} = \epsilon_{\dot\alpha\beta} = \begin{pmatrix} 0 & 1 \\ -1 & 0 \end{pmatrix}, \qquad \epsilon^{\alpha\gamma}\epsilon_{\beta\gamma} = \delta^{\alpha}_{\beta}\,, \qquad \epsilon^{\dot\alpha\gamma}\epsilon_{\beta\gamma} = \delta^{\dot\alpha}_{\beta}\,. \tag{2.3}$$

The indices of any 2-spinors λ_α, $\tilde\lambda_{\dot\alpha}$ will be raised as $\lambda^\alpha = \epsilon^{\alpha\beta}\lambda_\beta$, $\tilde\lambda^{\dot\alpha} = \epsilon^{\dot\alpha\beta}\tilde\lambda_{\dot\beta}$. We will also use the associated antisymmetric brackets: $\langle\lambda\,\kappa\rangle := \lambda^\alpha\,\kappa_\alpha$, $[\tilde\lambda\,\tilde\kappa] := \tilde\lambda^{\dot\alpha}\,\tilde\kappa_{\dot\alpha}$.

Let $\mathbb{S}_{\mathbb{C}}$ and $\tilde{\mathbb{S}}_{\mathbb{C}}$ be the trivial rank-2 complex vector bundles on $\mathbb{C}^4$. We call them the bundles of undotted and dotted spinors and use λ_α, $\tilde\lambda_{\dot\alpha}$ respectively as holomorphic coordinates on their $\mathbb{C}^2$ fibers. Denoting complex projective n-space by $\mathbb{P}^n$, the projective (undotted) spinor bundle $\mathbb{PS}_{\mathbb{C}} = \mathbb{C}^4 \times \mathbb{P}^1$ is obtained by projectivizing the fibers of $\mathbb{S}$. Taking $\mathbb{P}^3$ with homogeneous coordinates $Z^A = (\mu^{\dot\alpha}, \lambda_\alpha)$, the twistor correspondence is then summarized by a double fibration

$$\begin{array}{ccccc} & & \mathbb{C}^4 \times \mathbb{P}^1 & & \\ & \overset{p}{\swarrow} & & \overset{q}{\searrow} & \\ \mathbb{P}^3 & & & & \mathbb{C}^4 \end{array}$$

with the maps p, q being given by

$$\begin{aligned} p &: (x, \lambda) \mapsto Z^A = (x^{\beta\dot\alpha}\lambda_\beta, \lambda_\alpha)\,, \\ q &: (x, \lambda) \mapsto x\,. \end{aligned} \tag{2.4}$$

The image $\mathbb{PT} := p(\mathbb{PS}_{\mathbb{C}}) \subset \mathbb{P}^3$ is known as the twistor space of complexified Minkowski space [2, 3]. The map $\mu^{\dot\alpha} = x^{\alpha\dot\alpha}\lambda_\alpha$ is called an incidence relation. The image of p at fixed x is a rationally embedded copy of $\mathbb{P}^1$ that we will denote by X. This gives us Penrose's twistor correspondence

$$x \in \mathbb{C}^4 \quad \longleftrightarrow \quad X \simeq \mathbb{P}^1 \,:\, \mu^{\dot\alpha} = x^{\alpha\dot\alpha}\lambda_\alpha \tag{2.5}$$

between points in space-time and projective lines in twistor space.

The above correspondence can be specialized to the three real slices $\mathbb{R}^4$, $\mathbb{R}^{1,3}$ and $\mathbb{R}^{2,2}$ of varying signatures. To obtain Minkowski space $\mathbb{R}^{1,3}$ as a real slice of $\mathbb{C}^4$, we impose that x^a are real, or equivalently that $x^{\alpha\dot\alpha}$ is a hermitian matrix. This leads to a real slice on which (2.1) restricts as a real Minkowski metric. Its twistor space is then defined to be the image of the real projective spinor bundle $\mathbb{R}^{1,3} \times \mathbb{P}^1$ under the incidence relations map p. For a given real point $x \in \mathbb{R}^{1,3}$, any twistor Z^A on the corresponding projective line X satisfies

$$\operatorname{Im} \mu^{\dot\alpha}\bar\lambda_{\dot\alpha} = 2\,\mathrm{i}\,(x - x^\dagger)^{\alpha\dot\alpha}\lambda_\alpha\bar\lambda_{\dot\alpha} = 0\,, \tag{2.6}$$

where Im stands for imaginary part, and $\bar\lambda_{\dot\alpha}$ is the hermitian conjugate of λ_α. The set of all twistors satisfying (2.6) constitutes the twistor space of real Minkowski space. But unfortunately this is one condition and its solution set is a 5-real-dimensional manifold. As a result, we lose many of the powerful tools of complex analytic geometry when working directly on this 5d twistor space of $\mathbb{R}^{1,3}$.

The situation is improved by working in Euclidean signature. This case will be the main setting of the earlier chapters of this thesis and will be discussed in the next section. The real slice $\mathbb{R}^{2,2}$ of split signature also finds numerous applications in older studies of scattering amplitudes. We will only use it occasionally in examples like plane wave space-times, but the interested reader can find a detailed review of the associated twistor geometry in [4].

2.2 Euclidean Twistor Theory

The twistor theory of Riemannian 4-manifolds was developed in [5]; we follow the treatment reviewed in [6–8]. To obtain $\mathbb{R}^4$ with the Euclidean metric, we Wick rotate the spatial coordinates $x^i \mapsto \mathrm{i}\,x^i$, $i = 1, 2, 3$, and subsequently demand that all x^a be real. This converts (2.1) into the Euclidean metric $\mathrm{d}s^2 = (\mathrm{d}x^0)^2 + (\mathrm{d}x^1)^2 + (\mathrm{d}x^2)^2 + (\mathrm{d}x^3)^2$. The resulting spinor coordinates read

$$x^{\alpha\dot\alpha} = \frac{1}{\sqrt{2}}\begin{pmatrix} x^0 + \mathrm{i}\,x^3 & x^2 + \mathrm{i}\,x^1 \\ -x^2 + \mathrm{i}\,x^1 & x^0 - \mathrm{i}\,x^3 \end{pmatrix}, \tag{2.7}$$

in terms of which $\mathrm{d}s^2 = \mathrm{d}x^{\alpha\dot\alpha}\,\mathrm{d}x_{\alpha\dot\alpha} \equiv 2\,\det(\mathrm{d}x^{\alpha\dot\alpha})$.

The associated twistor space is the image of the real projective spinor bundle $\mathbb{PS} = \mathbb{R}^4 \times \mathbb{P}^1$ under the incidence relations map $p : (x, \lambda_\alpha) \mapsto Z^A = (x^{\beta\dot\alpha}\lambda_\beta, \lambda_\alpha)$. This is top-dimensional in $\mathbb{P}^3$ and can be checked to equal the full twistor space $\mathbb{PT}$ of complexified Minkowski space. It is explicitly given by

$$\mathbb{PT} = \{Z^A = (\mu^{\dot\alpha}, \lambda_\alpha) \in \mathbb{P}^3 \mid \lambda_\alpha \neq 0\}\,. \tag{2.8}$$

Here, we have noted that the λ_α cannot be zero simultaneously since they were homogeneous coordinates on $\mathbb{P}^1$ before applying p. As a complex manifold, $\mathbb{PT}$ can also be identified with the total space of the rank 2 holomorphic vector bundle $\mathcal{O}(1) \oplus \mathcal{O}(1) \to \mathbb{P}^1$, with λ_α and $\mu^{\dot\alpha}$ acting as holomorphic coordinates along the base and up the fibers respectively.

The map p is a diffeomorphism from $\mathbb{R}^4 \times \mathbb{P}^1$ to $\mathbb{PT}$. Indeed, complex conjugating $\mu^{\dot\alpha} = x^{\alpha\dot\alpha}\lambda_\alpha$ and using the reality of the x^a occurring in (2.7) shows that $\hat\mu^{\dot\alpha} = x^{\alpha\dot\alpha}\hat\lambda_\alpha$. Here, we have introduced the "quaternionic conjugates"

$$\hat\lambda_\alpha = (-\overline{\lambda_2}, \overline{\lambda_1})\,, \qquad \hat\mu^{\dot\alpha} = (-\overline{\mu^{\dot 2}}, \overline{\mu^{\dot 1}})\,, \qquad \hat Z^A = (\hat\mu^{\dot\alpha}, \hat\lambda_\alpha) \tag{2.9}$$

satisfying the useful properties $\langle\hat\lambda\,\lambda\rangle = |\lambda_1|^2 + |\lambda_2|^2 \equiv \|\lambda\|^2$ and $[\hat\mu\,\mu] = \|\mu\|^2$. Since giving two points Z^A and $\hat Z^A$ on a projective line X determines it uniquely, we can invert p to find

$$p^{-1} : Z^A = (\mu^{\dot\alpha}, \lambda_\alpha) \mapsto x^{\alpha\dot\alpha} = \frac{\hat\lambda^\alpha \mu^{\dot\alpha} - \lambda^\alpha \hat\mu^{\dot\alpha}}{\langle \hat\lambda \, \lambda \rangle} . \tag{2.10}$$

As a result, we can work with either holomorphic coordinates ($\mu^{\dot\alpha}$, λ_α), or simply identify $\mathbb{PT}$ with $\mathbb{PS}$ and work with non-holomorphic coordinates ($x^{\alpha\dot\alpha}$, λ_α). We will prefer to do the latter as it makes the relation with space-time manifest.

In the complex structure on $\mathbb{PT}$, the (0, 1)-vector fields are spanned by

$$\bar\partial_0 = \langle \hat\lambda \, \lambda \rangle \, \lambda_\alpha \frac{\partial}{\partial \hat\lambda_\alpha} , \qquad \bar\partial_{\dot\alpha} = \lambda^\alpha \, \partial_{\alpha\dot\alpha} , \tag{2.11}$$

where $\partial_{\alpha\dot\alpha} \equiv \partial/\partial x^{\alpha\dot\alpha}$. It is easily verified using $\mu^{\dot\alpha} = x^{\alpha\dot\alpha}\lambda_\alpha$ that these are in the span of $\{\partial/\partial\hat\lambda_\alpha, \partial/\partial\hat\mu^{\dot\alpha}\}$. Just like (2.11), all other vector fields and differential forms in what follows will be normalized so as to have homogeneity 0 in $\hat\lambda_\alpha$. To be precise, $\bar\partial_0$ and $\bar\partial_{\dot\alpha}$ are not vector fields on $\mathbb{PT}$ per se. Rather, they provide useful global presentations of the (0, 1)-vector fields as sections of $T^{0,1}\mathbb{PT} \otimes \mathcal{O}(2)$ and $T^{0,1}\mathbb{PT} \otimes \mathcal{O}(1)$ respectively. Here, $\mathcal{O}(n) \to \mathbb{PT}$ is the pullback of the line bundle $\mathcal{O}(n) \to ¶^1$ by the projection map $\mathbb{PT} \simeq \mathbb{R}^4 \times \mathbb{P}^1 \to \mathbb{P}^1$. Smooth sections of $\mathcal{O}(n)$ are functions of x, λ_α, $\hat\lambda_\alpha$ with homogeneity n in λ_α and 0 in $\hat\lambda_\alpha$.

The $(1, 0)$-vector fields are found by complex conjugation of (2.11) and are spanned by

$$\partial_0 = -\frac{\hat\lambda_\alpha}{\langle \hat\lambda \, \lambda \rangle} \frac{\partial}{\partial \lambda_\alpha} , \qquad \partial_{\dot\alpha} = \frac{\hat\lambda^\alpha \, \partial_{\alpha\dot\alpha}}{\langle \hat\lambda \, \lambda \rangle} . \tag{2.12}$$

These are valued in $\mathcal{O}(-2)$ and $\mathcal{O}(-1)$ respectively. In particular, $\partial_{\dot\alpha} = \partial/\partial\mu^{\dot\alpha}$ in the ($\mu^{\dot\alpha}$, λ_α) coordinates. Some useful identities involving these vector fields are

$$\partial_{\alpha\dot\alpha} = \lambda_\alpha \, \partial_{\dot\alpha} - \frac{\hat\lambda_\alpha \, \bar\partial_{\dot\alpha}}{\langle \hat\lambda \, \lambda \rangle} , \qquad [\bar\partial_0, \partial_{\dot\alpha}] = \bar\partial_{\dot\alpha} , \qquad [\bar\partial_{\dot\alpha}, \partial_0] = \partial_{\dot\alpha} . \tag{2.13}$$

We will also need the dual basis of (0, 1)-forms,

$$\bar e^0 = \frac{\mathrm{D}\hat\lambda}{\langle \hat\lambda \, \lambda \rangle^2} , \qquad \bar e^{\dot\alpha} = -\frac{\hat\lambda_\alpha \, \mathrm{d}x^{\alpha\dot\alpha}}{\langle \hat\lambda \, \lambda \rangle} , \tag{2.14}$$

where $\mathrm{D}\hat\lambda := \langle \hat\lambda \, \mathrm{d}\hat\lambda \rangle$, as well as the dual basis of (1, 0)-forms,

$$e^0 = \mathrm{D}\lambda \qquad e^{\dot\alpha} = \lambda_\alpha \, \mathrm{d}x^{\alpha\dot\alpha} , \tag{2.15}$$

where $\mathrm{D}\lambda := \langle \lambda \, \mathrm{d}\lambda \rangle$. The 1-form $\mathrm{D}\lambda$ trivializes the canonical bundle over $\mathbb{P}^1$ and identifies it with the line bundle $\mathcal{O}(-2)$. In the same vein, the bundle of spinors over $\mathbb{P}^1$ is given by its square root, the tautological bundle $\mathcal{O}(-1)$, an object that will often crop up in the study of sigma models.

Calculus on $\mathbb{PT}$. The set of smooth (p, q)-forms on $\mathbb{PT}$ valued in a complex vector bundle $E \to \mathbb{PT}$ will be denoted by $\Omega^{p,q}(\mathbb{PT}, E)$. Also set $\Omega^k(\mathbb{PT}, E) = \bigoplus_{p+q=k} \Omega^{p,q}(\mathbb{PT}, E)$ as usual. We will primarily encounter the case $E = \mathcal{O}(n)$. Letting

$$\Upsilon = \lambda_\alpha \frac{\partial}{\partial \lambda_\alpha}\,, \qquad \hat{\Upsilon} = \hat{\lambda}_\alpha \frac{\partial}{\partial \hat{\lambda}_\alpha} \tag{2.16}$$

be the holomorphic and anti-holomorphic Euler vector fields on $\mathbb{C}^2$, any differential form $\omega \in \Omega^{p,q}(\mathbb{PT}, \mathcal{O}(n))$ can be viewed as a form on the real non-projective spinor bundle $\mathbb{S} = \mathbb{R}^4 \times \mathbb{C}^2$ satisfying the homogeneity requirements

$$\mathcal{L}_\Upsilon \omega = n\,\omega\,, \qquad \mathcal{L}_{\hat{\Upsilon}} \omega = 0\,, \qquad \Upsilon \lrcorner\, \omega = 0\,, \qquad \hat{\Upsilon} \lrcorner\, \omega = 0\,, \tag{2.17}$$

where $\lrcorner$ signifies interior product, and $\mathcal{L}$ is the ordinary Lie derivative along $\mathbb{R}^4 \times \mathbb{C}^2$. Eg., the $(0, 1)$- and $(1, 0)$-forms $\bar{e}^0$, $\bar{e}^{\dot{\alpha}}$, e^0, $e^{\dot{\alpha}}$ are valued in $\mathcal{O}(-2)$, $\mathcal{O}(-1)$, $\mathcal{O}(2)$, $\mathcal{O}(1)$ respectively. Smooth $\mathcal{O}(n)$-valued vector fields on $\mathbb{PT}$ are analogously identified with vector fields V on $\mathbb{S}$ taken modulo Υ, $\hat{\Upsilon}$ and satisfying $\mathcal{L}_\Upsilon V = [\Upsilon, V] = n\,V$ along with $\mathcal{L}_{\hat{\Upsilon}} V = 0$.

Operations like the exterior and Lie derivatives along $\mathbb{R}^4 \times \mathbb{P}^1$ do not trivially coincide with their non-projective counterparts on $\mathbb{R}^4 \times \mathbb{C}^2$ even in homogeneous coordinates. This is because they typically do not map $\mathcal{O}(n)$-valued forms to $\mathcal{O}(n)$-valued forms. The remedy is to recall that $\mathcal{O}(n) \to \mathbb{P}^1$ is isomorphic to $(T^{1,0}\mathbb{P}^1)^{n/2}$. The Levi-Civita connection of the stereographic metric on $\mathbb{P}^1$ then lifts to a Chern connection on $\mathcal{O}(n)$:

$$\mathrm{d} = \mathrm{d}_{\mathbb{S}} - n\,\frac{\langle \hat{\lambda}\,\mathrm{d}\lambda\rangle}{\langle \hat{\lambda}\,\lambda\rangle} \wedge\,. \tag{2.18}$$

Here $\mathrm{d}_{\mathbb{S}} = \mathrm{d}\lambda_\alpha\,\partial_{\lambda_\alpha} + \mathrm{d}\hat{\lambda}_\alpha\,\partial_{\hat{\lambda}_\alpha} + \mathrm{d}x^{\alpha\dot{\alpha}}\,\partial_{\alpha\dot{\alpha}}$ is the exterior derivative on $\mathbb{S}$. And we are continuing to denote the exterior derivative of $\mathcal{O}(n)$-valued forms by d for ease of notation. The salient feature of (2.18) is that if ω satisfies (2.17), then $\mathrm{d}\omega$ also satisfies (2.17). Eg., acting with d on $\mathcal{O}(n)$-valued functions f is the same as replacing $\mathrm{d}\lambda_\alpha\,\partial_{\lambda_\alpha} f + \mathrm{d}\hat{\lambda}_\alpha\,\partial_{\hat{\lambda}_\alpha} f$ by $e^0\,\eth_0 f + \bar{e}^0\,\bar{\eth}_0 f$.

Similarly, we will define Lie derivatives of $\mathcal{O}(m)$-valued forms ω along $\mathcal{O}(l)$-valued vector fields V by using (2.18) for the cases $n = m, l + m$ as the exterior derivative in Cartan's homotopy formula $\mathcal{L}_V \omega = V \lrcorner\, \mathrm{d}\omega + \mathrm{d}(V \lrcorner\, \omega)$. The fact that $V \lrcorner\, \omega$ and $V \lrcorner\, \mathrm{d}\omega$ are valued in $\mathcal{O}(l + m)$ follows from the Leibniz rule for $\mathcal{L}_\Upsilon$ on $\mathbb{S}$. So $\mathcal{L}_V \omega$ is also valued in $\mathcal{O}(l + m)$.

Since twistor space is a complex manifold, we can also decompose d into holomorphic and anti-holomorphic exterior derivatives: $\mathrm{d} = \partial + \bar{\partial}$. Let $\pi_{r,s}$ be the map that projects a differential form ω onto the space of (r, s)-forms. Then ∂ and $\bar{\partial}$ are defined by setting

$$\partial\omega := \pi_{p+1,q}(\mathrm{d}\omega)\,, \qquad \bar{\partial}\omega := \pi_{p,q+1}(\mathrm{d}\omega) \qquad \forall\, \omega \in \Omega^{p,q}\,. \tag{2.19}$$

This prescription straightforwardly extends to $\mathcal{O}(n)$-valued forms by extending d via (2.18). The operator $\bar{\partial}$ is called a Dolbeault operator and will occur ubiquitously in building holomorphic theories on twistor space.

To evaluate ∂ and $\bar{\partial}$ in our bases (2.14) and (2.15), we need the structure equations:

$$\mathrm{d}e^0 = \mathrm{d}\bar{e}^0 = 0\,, \qquad \mathrm{d}e^{\dot{\alpha}} = e^0 \wedge \bar{e}^{\dot{\alpha}}\,, \qquad \mathrm{d}\bar{e}^{\dot{\alpha}} = e^{\dot{\alpha}} \wedge \bar{e}^0\,. \tag{2.20}$$

These are found by direct computation using (2.18). In particular, since $\mathrm{d}\bar{e}^0$ and $\mathrm{d}\bar{e}^{\dot{\alpha}}$ have no (0, 2)-form parts, we find that $\bar{\partial}\bar{e}^0 = \bar{\partial}\bar{e}^{\dot{\alpha}} = 0$, whereas $\partial\bar{e}^0 = 0$, $\partial\bar{e}^{\dot{\alpha}} = e^{\dot{\alpha}} \wedge \bar{e}^0$. As an application that we will frequently encounter, suppose ω is an $\mathcal{O}(n)$-valued (0, 1)-form

$$\omega = \omega_0\,\bar{e}^0 + \omega_{\dot{\alpha}}\,\bar{e}^{\dot{\alpha}}\,, \tag{2.21}$$

where $\omega_0 \in \Omega^0(\mathbb{PT}, \mathcal{O}(n+2))$ and $\omega_{\dot{\alpha}} \in \Omega^0(\mathbb{PT}, \mathcal{O}(n+1))$. Its anti-holomorphic derivative is found to be

$$\bar{\partial}\omega = \pi_{0,2}(\mathrm{d}\omega) = (\bar{\partial}_0\omega_{\dot{\alpha}} - \bar{\partial}_{\dot{\alpha}}\omega_0)\,\bar{e}^0 \wedge \bar{e}^{\dot{\alpha}} + \bar{\partial}_{\dot{\alpha}}\omega_{\dot{\beta}}\,\bar{e}^{\dot{\alpha}} \wedge \bar{e}^{\dot{\beta}}\,. \tag{2.22}$$

Similarly, we can also compute the holomorphic derivative

$$\partial\omega = \pi_{1,1}(\mathrm{d}\omega) = e^0 \wedge \left(\partial_0\omega_0\,\bar{e}^0 + \partial_0\omega_{\dot{\alpha}}\,\bar{e}^{\dot{\alpha}}\right) + e^{\dot{\alpha}} \wedge \left[(\partial_{\dot{\alpha}}\omega_0 + \omega_{\dot{\alpha}})\,\bar{e}^0 + \partial_{\dot{\alpha}}\omega_{\dot{\beta}}\,\bar{e}^{\dot{\beta}}\right]. \tag{2.23}$$

Among Lie derivatives, we will come across derivatives along $\partial_{\dot{\alpha}}$ in various Penrose integral formulae. Remembering $\mathrm{d}\omega = \partial\omega + \bar{\partial}\omega$, we can use Cartan's homotopy formula to find

$$\mathcal{L}_{\partial_{\dot{\alpha}}}\omega = (\partial_{\dot{\alpha}}\omega_0 + \omega_{\dot{\alpha}})\,\bar{e}^0 + \partial_{\dot{\alpha}}\omega_{\dot{\beta}}\,\bar{e}^{\dot{\beta}}\,. \tag{2.24}$$

Notice that the $\bar{e}^0$-component contains an extra term $\omega_{\dot{\alpha}}$ that can be traced back to the last equation in (2.20).

2.3 Penrose Transform for Linear Fields

Twistor theory can be used to find solutions to linearized massless field equations on flat space [9]. A classic example of this is the Penrose contour integral formula. Let $x \in \mathbb{C}^4$ and let $X \simeq \mathbb{P}^1$ be its corresponding twistor line $\mu^{\dot{\alpha}} = x^{\alpha\dot{\alpha}}\lambda_\alpha$. For any meromorphic function $f \in \Omega^0(\mathbb{PT}, \mathcal{O}(-2))$, its contour integral around one of its poles,

$$\varphi(x) = \frac{1}{2\pi\mathrm{i}}\oint \mathrm{D}\lambda\; f\big|_X\,, \tag{2.25}$$

solves the Klein-Gordon equation $\Box\varphi = 0$ on complexified Minkowski space. Similar formulae also exist for higher spin massless free fields. We now review more convenient versions of such integral formulae which trade contour integrals for integrals over the entire twistor line X [6]. We will find it quickest to prove them on a Euclidean real slice $\mathbb{R}^4 \subset \mathbb{C}^4$.

Generalizing the scalar field, a massless negative helicity field of helicity $\ell < 0$ is represented by a field strength $\varphi_{\alpha_1\alpha_2\cdots\alpha_{2|\ell|}}(x)$ symmetrized in its spinor indices. It is on-shell if it satisfies the linearized field equation [10]

$$\partial^{\alpha_1\dot{\alpha}_1}\varphi_{\alpha_1\alpha_2\cdots\alpha_{2|\ell|}} = 0\,. \tag{2.26}$$

The conjugate positive helicity fields with helicity $\ell > 0$ are given by field strengths $\varphi_{\dot{\alpha}_1\dot{\alpha}_2\cdots\dot{\alpha}_{2\ell}}(x)$ with symmetrized dotted spinor indices. They satisfy the field equations

$$\partial^{\alpha_1\dot{\alpha}_1}\varphi_{\dot{\alpha}_1\dot{\alpha}_2\cdots\dot{\alpha}_{2\ell}} = 0\,. \tag{2.27}$$

According to the twistor correspondence, massless free fields of helicity ℓ correspond to elements of the first Dolbeault cohomology groups $H^{0,1}(\mathbb{PT}, \mathcal{O}(2\ell-2))$ on twistor space.

The case of negative helicity $\ell < 0$ is relatively simpler. Let $\omega \in H^{0,1}(\mathbb{PT}, \mathcal{O}(2\ell - 2))$, so that it satisfies $\bar{\partial}\omega = 0$. A representative of the cohomology class of ω will be referred to as a twistor representative. Define a negative helicity space-time field by the integral formula

$$\varphi_{\alpha_1\alpha_2\cdots\alpha_{2|\ell|}}(x) = \int_X \mathrm{D}\lambda \wedge \lambda_{\alpha_1}\lambda_{\alpha_2}\cdots\lambda_{\alpha_{2|\ell|}}\,\omega\big|_X\,. \tag{2.28}$$

The integrand on the right has zero homogeneity in λ_α, so that the integral over $X \simeq \mathbb{P}^1$ is well-defined. Since λ_α and $\mathrm{D}\lambda = \langle\lambda\,\mathrm{d}\lambda\rangle$ are holomorphic, the integral is invariant under shifts $\omega \mapsto \omega + \bar{\partial}\chi$ for $\chi \in \Omega^0(\mathbb{PT}, \mathcal{O}(2\ell - 2))$. Writing $\omega = \omega_0\,\bar{e}^0 + \omega_{\dot{\alpha}}\,\bar{e}^{\dot{\alpha}}$ in our basis of (0, 1)-forms, we see that $\omega|_X = \omega_0(x, \lambda, \hat{\lambda})\,\bar{e}^0$. To show that this satisfies (2.26), we compute

$$\partial^{\alpha_1\dot{\alpha}_1}\varphi_{\alpha_1\alpha_2\cdots\alpha_{2|\ell|}} = -\int_X \mathrm{D}\lambda \wedge \lambda_{\alpha_2}\cdots\lambda_{\alpha_{2|\ell|}}\,\bar{\partial}^{\dot{\alpha}_1}\omega_0|_X\,\bar{e}^0\,, \tag{2.29}$$

having used $\bar{\partial}_{\dot{\alpha}} = \lambda^\alpha\,\partial_{\alpha\dot{\alpha}}$. Since $\bar{\partial}\omega = 0$, it follows from (2.22) that $\bar{\partial}_0\omega_{\dot{\alpha}} = \bar{\partial}_{\dot{\alpha}}\omega_0$ and $\bar{\partial}_{\dot{\alpha}}\omega_{\dot{\beta}} = \bar{\partial}_{\dot{\beta}}\omega_{\dot{\alpha}}$. We can use the first of these to rewrite (2.29) as

$$\begin{aligned}\partial^{\alpha_1\dot{\alpha}_1}\varphi_{\alpha_1\alpha_2\cdots\alpha_{2|\ell|}} &= -\int_X \mathrm{D}\lambda \wedge \lambda_{\alpha_2}\cdots\lambda_{\alpha_{2|\ell|}}\,\bar{\partial}_0\omega^{\dot{\alpha}_1}|_X\,\bar{e}^0 \\ &= -\int_X \mathrm{D}\lambda \wedge \bar{\partial}|_X\big(\lambda_{\alpha_2}\cdots\lambda_{\alpha_{2|\ell|}}\,\omega^{\dot{\alpha}_1}|_X\big) = 0\,,\end{aligned} \tag{2.30}$$

written in terms of the Dolbeault operator $\bar{\partial}|_X \equiv \bar{e}^0\,\bar{\partial}_0$ of $\mathbb{P}^1$. The final vanishing followed from Stokes' theorem.

For a scalar field $\varphi(x)$, we simply write a surface integral analogue of (2.25):

$$\varphi(x) = \int_X \mathrm{D}\lambda \wedge \omega\big|_X \,, \tag{2.31}$$

where $\omega \in H^{0,1}(\mathbb{PT}, \mathcal{O}(-2))$. Acting with the Laplacian $\Box = \partial^{\alpha\dot{\alpha}}\partial_{\alpha\dot{\alpha}}$ yields

$$\Box\varphi(x) = \int_X \mathrm{D}\lambda \wedge \partial^{\alpha\dot{\alpha}}\partial_{\alpha\dot{\alpha}}\omega_0|_X\,\bar{e}^0 = 2\int_X \mathrm{D}\lambda \wedge \partial^{\dot{\alpha}}\bar{\partial}_{\dot{\alpha}}\omega_0|_X\,\bar{e}^0 \,, \tag{2.32}$$

having used the first identity in (2.13). Further simplifying using $\bar{\partial}_{\dot{\alpha}}\omega_0 = \bar{\partial}_0\omega_{\dot{\alpha}}$ then gives

$$\Box\varphi(x) = 2\int_X \mathrm{D}\lambda \wedge \partial^{\dot{\alpha}}\bar{\partial}_0\omega_{\dot{\alpha}}|_X\,\bar{e}^0 = 2\int_X \mathrm{D}\lambda \wedge \left(\bar{\partial}_0\partial^{\dot{\alpha}}\omega_{\dot{\alpha}}|_X - \bar{\partial}^{\dot{\alpha}}\omega_{\dot{\alpha}}|_X\right)\bar{e}^0 = 0\,. \tag{2.33}$$

To get the second equality, we used $[\bar{\partial}_0, \partial_{\dot{\alpha}}] = \bar{\partial}_{\dot{\alpha}}$. For the final equality, we again used Stokes' theorem and $\bar{\partial}$-closure $\bar{\partial}^{\dot{\alpha}}\omega_{\dot{\alpha}} = \epsilon^{\dot{\beta}\dot{\alpha}}\,\bar{\partial}_{[\dot{\alpha}}\omega_{\dot{\beta}]} = 0$.

The integral formulae for $\ell > 0$ are slightly more involved. For $\omega \in H^{0,1}(\mathbb{PT}, \mathcal{O}(2\ell-2))$, the corresponding positive helicity field strength can be expressed as

$$\varphi_{\dot{\alpha}_1\dot{\alpha}_2\cdots\dot{\alpha}_{2\ell}}(x) = \int_X \mathrm{D}\lambda \wedge \mathcal{L}_{\partial_{\dot{\alpha}_1}}\mathcal{L}_{\partial_{\dot{\alpha}_2}}\cdots\mathcal{L}_{\partial_{\dot{\alpha}_{2\ell}}}\omega\,\big|_X\,. \tag{2.34}$$

The Lie derivatives of ω are to be evaluated before restricting to X. They can be evaluated inductively using (2.24),

$$\mathcal{L}_{\partial_{\dot{\alpha}_1}}\mathcal{L}_{\partial_{\dot{\alpha}_2}}\cdots\mathcal{L}_{\partial_{\dot{\alpha}_{2\ell}}}\omega\,\big|_X = \left(\partial_{\dot{\alpha}_1}\partial_{\dot{\alpha}_2}\cdots\partial_{\dot{\alpha}_{2\ell}}\omega_0 + 2\ell\,\partial_{(\dot{\alpha}_2}\cdots\partial_{\dot{\alpha}_{2\ell}}\omega_{\dot{\alpha}_1)}\right)\big|_X\,\bar{e}^0\,. \tag{2.35}$$

To verify the field Eq. (2.27), we compute

$$\begin{aligned}
&\partial^{\alpha_1\dot{\alpha}_1}\left(\partial_{\dot{\alpha}_1}\partial_{\dot{\alpha}_2}\cdots\partial_{\dot{\alpha}_{2\ell}}\omega_0 + 2\ell\,\partial_{(\dot{\alpha}_2}\cdots\partial_{\dot{\alpha}_{2\ell}}\omega_{\dot{\alpha}_1)}\right)\\
&= \lambda^{\alpha_1}\partial^{\dot{\alpha}_1}\partial_{\dot{\alpha}_2}\cdots\partial_{\dot{\alpha}_{2\ell}}\omega_{\dot{\alpha}_1} - \frac{\hat{\lambda}^{\alpha_1}}{\langle\hat{\lambda}\,\lambda\rangle}\left[\partial_{\dot{\alpha}_1}\partial_{\dot{\alpha}_2}\cdots\partial_{\dot{\alpha}_{2\ell}}\bar{\partial}^{\dot{\alpha}_1}\omega_0 + 2\ell\,\bar{\partial}^{\dot{\alpha}_1}\partial_{(\dot{\alpha}_2}\cdots\partial_{\dot{\alpha}_{2\ell}}\omega_{\dot{\alpha}_1)}\right]\\
&= \lambda^{\alpha_1}\partial^{\dot{\alpha}_1}\partial_{\dot{\alpha}_2}\cdots\partial_{\dot{\alpha}_{2\ell}}\omega_{\dot{\alpha}_1} - \frac{\hat{\lambda}^{\alpha_1}}{\langle\hat{\lambda}\,\lambda\rangle}\left[\partial_{\dot{\alpha}_1}\partial_{\dot{\alpha}_2}\cdots\partial_{\dot{\alpha}_{2\ell}}\bar{\partial}_0\omega^{\dot{\alpha}_1} + (2\ell-1)\,\partial_{\dot{\alpha}_1}\bar{\partial}_{(\dot{\alpha}_2}\partial_{\dot{\alpha}_3}\cdots\partial_{\dot{\alpha}_{2\ell})}\omega^{\dot{\alpha}_1}\right]\\
&= \bar{\partial}_0\left(\frac{\hat{\lambda}^{\alpha_1}}{\langle\hat{\lambda}\,\lambda\rangle}\,\partial^{\dot{\alpha}_1}\partial_{\dot{\alpha}_2}\cdots\partial_{\dot{\alpha}_{2\ell}}\omega_{\dot{\alpha}_1}\right).
\end{aligned} \tag{2.36}$$

The second line follows by applying the first of (2.13) and dropping terms containing $\partial^{\dot{\alpha}_1}\partial_{\dot{\alpha}_1} = 0$ due to antisymmetry. The third line is a consequence of the $\bar{\partial}$-closure conditions $\bar{\partial}^{\dot{\alpha}_1}\omega_0 = \bar{\partial}_0\omega^{\dot{\alpha}_1}$ and $\bar{\partial}^{\dot{\alpha}_1}\omega_{\dot{\alpha}_k} = \bar{\partial}_{\dot{\alpha}_k}\omega^{\dot{\alpha}_1}$. Equality of the third and fourth lines can be checked by direct computation using the commutator $[\bar{\partial}_0, \partial_{\dot{\alpha}}] = \bar{\partial}_{\dot{\alpha}}$. Stokes' theorem finally allows us to conclude that

$$\partial^{\alpha_1\dot{\alpha}_1}\varphi_{\dot{\alpha}_1\dot{\alpha}_2\cdots\dot{\alpha}_{2\ell}}(x) = \int_X \mathrm{D}\lambda \wedge \bar{\partial}|_X\left(\frac{\hat{\lambda}^{\alpha_1}}{\langle\hat{\lambda}\,\lambda\rangle}\,\partial^{\dot{\alpha}_1}\partial_{\dot{\alpha}_2}\cdots\partial_{\dot{\alpha}_{2\ell}}\omega_{\dot{\alpha}_1}\big|_X\right) = 0\,, \tag{2.37}$$

giving us a solution to the positive helicity field equations.

With these formulae, we can explicitly construct on-shell massless fields on $\mathbb{R}^4$ from $\bar{\partial}$-closed (0, 1)-forms on $\mathbb{PT}$. In many situations, one would like to invert this procedure. This can also be done fairly explicitly. Up to addition of $\bar{\partial}$-exact terms, the Euclidean signature twistor representative for a scalar or negative helicity field $\varphi_{\alpha_1\cdots\alpha_{2|\ell|}}(x)$ can be taken to be

$$\omega = (2|\ell|+1)\,\varphi_{\alpha_1\cdots\alpha_{2|\ell|}}\,\frac{\hat{\lambda}^{\alpha_1}\ldots\hat{\lambda}^{\alpha_{2|\ell|}}}{\langle\hat{\lambda}\,\lambda\rangle^{2|\ell|}}\,\bar{e}^0 + \partial_{\alpha_1\dot{\beta}}\varphi_{\alpha_2\cdots\alpha_{2|\ell|+1}}\,\frac{\hat{\lambda}^{\alpha_1}\ldots\hat{\lambda}^{\alpha_{2|\ell|+1}}}{\langle\hat{\lambda}\,\lambda\rangle^{2|\ell|+1}}\,\bar{e}^{\dot{\beta}}\,. \tag{2.38}$$

A direct calculation using (2.22) establishes that $\bar{\partial}\omega = 0$ if and only if $\partial^{\alpha_1\dot{\alpha}_1}\varphi_{\alpha_1\cdots\alpha_{2|\ell|}} = 0$.

On the other hand, twistor representatives for positive helicity fields are generally written in terms of potentials for the field strengths $\varphi_{\dot{\alpha}_1\dot{\alpha}_2\cdots\dot{\alpha}_{2\ell}}(x)$ [10]. Suppose $\psi_{\alpha_1\alpha_2\cdots\alpha_{2\ell-1}\dot{\alpha}_{2\ell}}(x)$ locally acts as such a potential, in the sense that it is symmetric in its undotted indices and gives rise to the field strength

$$\varphi_{\dot{\alpha}_1\dot{\alpha}_2\cdots\dot{\alpha}_{2\ell}} = \partial^{\alpha_1}_{(\dot{\alpha}_1}\partial^{\alpha_2}_{\dot{\alpha}_2}\cdots\partial^{\alpha_{2\ell-1}}_{\dot{\alpha}_{2\ell-1}|}\psi_{\alpha_1\alpha_2\cdots\alpha_{2\ell-1}|\dot{\alpha}_{2\ell})}\,. \tag{2.39}$$

The field Eq. (2.27) for the field strengths emerge from repeated differentiation of field equations for the potential:

$$\partial^{\dot{\alpha}_{2\ell}}_{(\alpha_{2\ell}}\psi_{\alpha_1\alpha_2\cdots\alpha_{2\ell-1})\dot{\alpha}_{2\ell}} = 0\,. \tag{2.40}$$

The associated gauge symmetry is $\psi_{\alpha_1\alpha_2\cdots\alpha_{2\ell-1}\dot{\alpha}_{2\ell}} \sim \psi_{\alpha_1\alpha_2\cdots\alpha_{2\ell-1}\dot{\alpha}_{2\ell}} + \partial_{(\alpha_1|\dot{\alpha}_{2\ell}}\chi_{|\alpha_2\cdots\alpha_{2\ell-1})}$.

For any $\omega \in H^{0,1}(\mathbb{PT}, \mathcal{O}(2\ell-2))$ and a choice of constant "reference spinor" ξ_α, the field Eq. (2.40) are solved by Penrose integral formulae of the form

$$\psi_{\alpha_1\alpha_2\cdots\alpha_{2\ell-1}\dot{\alpha}_{2\ell}}(x) = \int_X \frac{\mathrm{D}\lambda}{\langle\lambda\,\xi\rangle^{2\ell-1}} \wedge \xi_{\alpha_1}\,\xi_{\alpha_2}\cdots\xi_{\alpha_{2\ell-1}}\,\mathcal{L}_{\partial_{\dot{\alpha}_{2\ell}}}\omega\,\big|_X \tag{2.41}$$

as long as ω vanishes to the $(2\ell-1)$th order at $\langle\lambda\,\xi\rangle = 0$. The choice of ξ_α acts as a "light-cone gauge" wherein we take the potential to satisfy $\xi^{\alpha_1}\psi_{\alpha_1\cdots\alpha_{2\ell-1}\dot{\alpha}_{2\ell}} = 0$. Substituting (2.41) into (2.39) and simplifying using $\bar{\partial}\omega = 0$ gives rise to the integral formulae (2.34) for the field strengths. Conversely, if we start with a potential and write

$$\omega = \lambda^{\alpha_1}\,\lambda^{\alpha_2}\ldots\lambda^{\alpha_{2\ell-1}}\,\psi_{\alpha_1\alpha_2\cdots\alpha_{2\ell-1}\dot{\beta}}\,\bar{e}^{\dot{\beta}}\,, \tag{2.42}$$

one sees that $\bar{\partial}\omega = 0$ if and only if the potential satisfies its field Eq. (2.40). If the potential is in light-cone gauge, (2.42) clearly vanishes to the required order at $\langle\lambda\,\xi\rangle = 0$.

References

1. Penrose R, Rindler W (1984) Spinors and space-time, vol 1. Cambridge monographs on mathematical physics. Cambridge University Press, Cambridge, UK
2. Penrose R (1967) Twistor algebra. J Math Phys 8:345
3. Penrose R, Rindler W (1986) Spinors and space-time, vol 2. Cambridge Monographs on Mathematical Physics. Cambridge University Press, Cambridge, UK
4. Mason LJ (2006) Global anti-self-dual Yang-Mills fields in split signature and their scattering. Journal für die reine und angewandte Mathematik 2006(597):105–133. arXiv:math-ph/0505039
5. Atiyah MF, Hitchin NJ, Singer IM (1978) Self-duality in four-dimensional Riemannian geometry. Proc R Soc Lond Ser A 362:425–461
6. Woodhouse NMJ (1985) Real methods in twistor theory. Class. Quantum Gravity 2:257–291
7. Mason LJ (2005) Twistor actions for non-self-dual fields: A New Foundation for twistor-string theory. JHEP 10:009. arXiv:hep-th/0507269
8. Jiang W (2008) Aspects of Yang-Mills theory in twistor space. PhD thesis, University of Oxford. arXiv:0809.0328
9. Penrose R (1969) Solutions of the zero-rest-mass equations. J Math Phys 10:38–39
10. Penrose R (1965) Zero rest mass fields including gravitation: asymptotic behavior. Proc R Soc Lond A284:159
11. He T, Lysov V, Mitra P, Strominger A (2015) BMS supertranslations and Weinberg's soft graviton theorem. JHEP 05:151. arXiv:1401.7026

Chapter 3
Sigma Models and Hyperkähler Geometry

The Penrose integral formulae sure seem like an awful amount of work to solve something as innocuous as the Laplace equation on $\mathbb{R}^4$. In this chapter, we will see that solutions to even non-linear, integrable field equations can be built from holomorphic data on twistor space. We first review Penrose's non-linear graviton construction that relates complex structure deformations of $\mathbb{PT}$ to self-dual Ricci-flat space-times. The metrics on these space-times are necessarily hyperkähler. We then explain how the associated Kähler and Plebanski scalar potentials can be encoded into certain two-dimensional sigma models governing holomorphic curves in deformed twistor spaces.

3.1 Self-duality and Hyperkähler Manifolds

For simplicity, we start with working in Euclidean signature; but our results will easily generalize to complexified space-times. To this end, let M be an oriented Riemannian 4-manifold with metric g. Let $\boldsymbol{\theta}^a = (\boldsymbol{\theta}^0, \boldsymbol{\theta}^1, \boldsymbol{\theta}^2, \boldsymbol{\theta}^3)$ be a tetrad for the metric, i.e., a frame for the cotangent bundle that diagonalizes the metric: $g = \delta_{ab}\, \boldsymbol{\theta}^a\, \boldsymbol{\theta}^b$. We can define a null-tetrad for g by using the same construction as we did for (2.7),

$$g = \epsilon_{\alpha\beta}\, \epsilon_{\dot\alpha\dot\beta}\, \boldsymbol{\theta}^{\alpha\dot\alpha}\, \boldsymbol{\theta}^{\beta\dot\beta} = 2\, \det(\boldsymbol{\theta}^{\alpha\dot\alpha})\,, \tag{3.1}$$

$$\boldsymbol{\theta}^{\alpha\dot\alpha} := \frac{1}{\sqrt{2}} \begin{pmatrix} \boldsymbol{\theta}^0 + \mathrm{i}\,\boldsymbol{\theta}^3 & \boldsymbol{\theta}^2 + \mathrm{i}\,\boldsymbol{\theta}^1 \\ -\boldsymbol{\theta}^2 + \mathrm{i}\,\boldsymbol{\theta}^1 & \boldsymbol{\theta}^0 - \mathrm{i}\,\boldsymbol{\theta}^3 \end{pmatrix}. \tag{3.2}$$

We will generically assume that M has a spin structure, so that the indices $\alpha = 1, 2$, $\dot\alpha = \dot 1, \dot 2$ can be treated as spinor indices of SU(2) and be raised or lowered with $\epsilon^{\alpha\beta}$, $\epsilon^{\dot\alpha\dot\beta}$ as before. Let $\mathbf{e}_{\alpha\dot\alpha}$ denote the dual tetrad, i.e., a frame of the tangent bundle of M that satisfies $\mathbf{e}_{\alpha\dot\alpha} \lrcorner\, \boldsymbol{\theta}^{\beta\dot\beta} = \delta_\alpha^\beta\, \delta_{\dot\alpha}^{\dot\beta}$. Spinor components of any tensor are then defined as its components in the frames $\{\boldsymbol{\theta}^{\alpha\dot\alpha}, \mathbf{e}_{\alpha\dot\alpha}\}$. Similarly, we will denote the Levi-Civita connection of g in this frame by $\nabla_{\alpha\dot\alpha}$.

A. Sharma, *Twistor Sigma Models*, Springer Theses,
https://doi.org/10.1007/978-3-031-50751-9_3

Let $*$ denote Hodge star with respect to g. Recall that it maps 2-forms to 2-forms in four dimensions. Since $*^2 = 1$, it splits the space of 2-forms on M into two 3-dimensional eigenspaces with eigenvalues ± 1. A 2-form F on M is said to be self-dual (SD) if it satisfies $*F = F$ and anti-self-dual (ASD) if $*F = -F$. Bases of ASD and SD 2-forms are respectively given by

$$\Sigma^{\alpha\beta} \equiv \Sigma^{(\alpha\beta)} = \theta^{\alpha\dot{\alpha}} \wedge \theta^{\beta}{}_{\dot{\alpha}}\,, \qquad \tilde{\Sigma}^{\dot{\alpha}\dot{\beta}} \equiv \tilde{\Sigma}^{(\dot{\alpha}\dot{\beta})} = \theta^{\alpha\dot{\alpha}} \wedge \theta_{\alpha}{}^{\dot{\beta}}\,. \tag{3.3}$$

Similarly, we can also introduce the ASD and SD spin connection 1-forms $\Gamma_{\alpha\beta} \equiv \Gamma_{(\alpha\beta)}$ and $\tilde{\Gamma}_{\dot{\alpha}\dot{\beta}} \equiv \tilde{\Gamma}_{(\dot{\alpha}\dot{\beta})}$ respectively. These are defined by Cartan's first structure equation,

$$\mathrm{d}\theta^{\alpha\dot{\alpha}} = \Gamma^{\alpha}{}_{\beta} \wedge \theta^{\beta\dot{\alpha}} + \tilde{\Gamma}^{\dot{\alpha}}{}_{\dot{\beta}} \wedge \theta^{\alpha\dot{\beta}}\,, \tag{3.4}$$

which is equivalent to the following structure equations for the 2-forms:

$$\mathrm{d}\Sigma^{\alpha\beta} = 2\,\Gamma^{(\alpha}{}_{\gamma} \wedge \Sigma^{\beta)\gamma}\,, \qquad \mathrm{d}\tilde{\Sigma}^{(\dot{\alpha}\dot{\beta})} = 2\,\tilde{\Gamma}^{(\dot{\alpha}}{}_{\dot{\gamma}} \wedge \tilde{\Sigma}^{\dot{\beta})\dot{\gamma}}\,. \tag{3.5}$$

The associated (linearized) gauge transformations are given by $\delta\Gamma_{\alpha\beta} = \mathrm{d}\chi_{\alpha\beta} + 2\,\Gamma_{(\alpha}{}^{\gamma}\,\chi_{\beta)\gamma}$ and $\delta\tilde{\Gamma}_{\dot{\alpha}\dot{\beta}} = \mathrm{d}\chi_{\dot{\alpha}\dot{\beta}} + 2\,\tilde{\Gamma}_{(\dot{\alpha}}{}^{\dot{\gamma}}\,\chi_{\dot{\beta})\dot{\gamma}}$ for symmetric scalars $\chi_{\alpha\beta} = \chi_{(\alpha\beta)}$, $\chi_{\dot{\alpha}\dot{\beta}} = \chi_{(\dot{\alpha}\dot{\beta})}$.

The curvatures of $\Gamma_{\alpha\beta}$, $\tilde{\Gamma}_{\dot{\alpha}\dot{\beta}}$ give rise to ASD and SD parts of the Riemann curvature 2-form,

$$R_{\alpha\beta} = \mathrm{d}\Gamma_{\alpha\beta} + \Gamma_{\alpha}{}^{\gamma} \wedge \Gamma_{\gamma\beta}\,, \qquad \tilde{R}_{\dot{\alpha}\dot{\beta}} = \mathrm{d}\tilde{\Gamma}_{\dot{\alpha}\dot{\beta}} + \tilde{\Gamma}_{\dot{\alpha}}{}^{\dot{\gamma}} \wedge \tilde{\Gamma}_{\dot{\gamma}\dot{\beta}}\,, \tag{3.6}$$

with the full Riemann tensor being encoded in $R_{\alpha\dot{\alpha}\beta\dot{\beta}} = \epsilon_{\dot{\alpha}\dot{\beta}}\,R_{\alpha\beta} + \epsilon_{\alpha\beta}\,\tilde{R}_{\dot{\alpha}\dot{\beta}}$. The curvature 2-forms admit well-known decompositions into ASD and SD parts:

$$R_{\alpha\beta} = \Psi_{\alpha\beta\gamma\delta}\,\Sigma^{\gamma\delta} + \Phi_{\alpha\beta\dot{\alpha}\dot{\beta}}\,\tilde{\Sigma}^{\dot{\alpha}\dot{\beta}} + \frac{R}{12}\,\Sigma_{\alpha\beta}\,, \tag{3.7}$$

$$\tilde{R}_{\dot{\alpha}\dot{\beta}} = \tilde{\Psi}_{\dot{\alpha}\dot{\beta}\dot{\gamma}\dot{\delta}}\,\tilde{\Sigma}^{\dot{\gamma}\dot{\delta}} + \Phi_{\alpha\beta\dot{\alpha}\dot{\beta}}\,\Sigma^{\alpha\beta} + \frac{R}{12}\,\tilde{\Sigma}_{\dot{\alpha}\dot{\beta}}\,. \tag{3.8}$$

Here, R is the Ricci scalar, and $\Phi_{\alpha\beta\dot{\alpha}\dot{\beta}}$ are spinor components of the Ricci tensor's trace-free part. The remaining fields $\Psi_{\alpha\beta\gamma\delta}$ and $\tilde{\Psi}_{\dot{\alpha}\dot{\beta}\dot{\gamma}\dot{\delta}}$ are the ASD and SD parts of the Weyl tensor and are known as Weyl spinors.

The 4-manifold (M, g) is said to be [1]

- **self-dual** if the ASD Weyl spinor vanishes: $\Psi_{\alpha\beta\gamma\delta} = 0$,
- **vacuum**, i.e., Ricci-flat if the Ricci tensor vanishes: $\Phi_{\alpha\beta\dot{\alpha}\dot{\beta}} = R = 0$, and
- **self-dual vacuum** if $R_{\alpha\beta} = 0$.

An SD vacuum 4-manifold is also colloquially known as being "half-flat". Unlike the Einstein vacuum equations, the self-duality equation $\Psi_{\alpha\beta\gamma\delta} = 0$ or the SD vacuum equation $R_{\alpha\beta} = 0$ are classically integrable and admit beautiful twistorial descriptions. Moreover, in four dimensions, the condition of being self-dual as well as Ricci-flat happens to be equivalent to M being hyperkähler. In the rest of this section, we provide a brief review of this fact and establish our conventions.

Hyperkähler geometry in four dimensions. When M is self-dual vacuum, the Riemann curvature 2-form $R_{\alpha\dot{\alpha}\beta\dot{\beta}} = \epsilon_{\alpha\beta}\,\tilde{R}_{\dot{\alpha}\dot{\beta}}$ has holonomy SU(2). Such a manifold is said to be hyperkähler. It comes equipped with a triplet of complex structures and Kähler forms that are simultaneously compatible with the metric g. In terms of the ASD 2-forms $\Sigma^{\alpha\beta}$, the three Kähler forms can be taken to be Σ^{12} and $\Sigma^{11} \pm \mathrm{i}\,\Sigma^{22}$. They are closed by virtue of the first of (3.5) because the ASD spin connection is flat and can be set to $\Gamma_{\alpha\beta} = 0$ up to gauge transformations.

Using $\mathrm{d}\Sigma^{\alpha\beta} = 0$, one can find local complex coordinates $(y^{\dot{\alpha}}, \tilde{y}^{\dot{\alpha}})$ on M so that

$$\Sigma^{11} = \mathrm{d}y^{\dot{\alpha}} \wedge \mathrm{d}y_{\dot{\alpha}}\,, \qquad \Sigma^{12} = -\,\Omega_{\dot{\alpha}\dot{\beta}}\,\mathrm{d}y^{\dot{\alpha}} \wedge \mathrm{d}\tilde{y}^{\dot{\beta}}\,, \qquad \Sigma^{22} = \mathrm{d}\tilde{y}^{\dot{\alpha}} \wedge \mathrm{d}\tilde{y}_{\dot{\alpha}}\,, \tag{3.9}$$

where the form of Σ^{11} and Σ^{22} follows from Darboux's theorem and their rank. Depending on our spin frames, $\tilde{y}^{\dot{\alpha}}$ will generally be a linear combination of $y^{\dot{\alpha}} = (y^{\dot{1}}, y^{\dot{2}})$ and its conjugate $\hat{y}^{\dot{\alpha}} = (-\overline{y^{\dot{2}}}, \overline{y^{\dot{1}}})$. The closure of Σ^{12} implies that

$$\Omega_{\dot{\alpha}\dot{\beta}} = \frac{\partial^2\Omega}{\partial y^{\dot{\alpha}}\partial\tilde{y}^{\dot{\beta}}}\,, \tag{3.10}$$

by the usual argument for the existence of a symplectic potential.

In what follows, we will abbreviate functions $f(y, \tilde{y})$ of $y^{\dot{\alpha}}, \tilde{y}^{\dot{\alpha}}$ as $f(y)$. The Kähler potential $\Omega(y)$ is often referred to as Plebanski's first fundamental form for the hyperkähler metric on M,

$$g = 2\,\Omega_{\dot{\alpha}\dot{\beta}}\,\mathrm{d}y^{\dot{\alpha}}\,\mathrm{d}\tilde{y}^{\dot{\beta}}\,. \tag{3.11}$$

Up to local SL(2, $\mathbb{C}$)-valued rotations of the dotted indices, the simplest choice of frame $\boldsymbol{\theta}^{\alpha\dot{\alpha}}$ for the cotangent bundle is

$$\boldsymbol{\theta}^{1\dot{\alpha}} = \mathrm{d}y^{\dot{\alpha}}\,, \qquad \boldsymbol{\theta}^{2\dot{\alpha}} = -\,\Omega^{\dot{\alpha}}{}_{\dot{\beta}}\,\mathrm{d}\tilde{y}^{\dot{\beta}}\,, \tag{3.12}$$

in terms of which the metric reads $g = \epsilon_{\alpha\beta}\,\epsilon_{\dot{\alpha}\dot{\beta}}\,\boldsymbol{\theta}^{\alpha\dot{\alpha}}\,\boldsymbol{\theta}^{\beta\dot{\beta}}$. It is a Ricci-flat metric if Ω satisfies

$$\det(\Omega_{\dot{\alpha}\dot{\beta}}) \equiv \frac{1}{2}\,\frac{\partial^2\Omega}{\partial y^{\dot{\alpha}}\partial\tilde{y}^{\dot{\beta}}}\,\frac{\partial^2\Omega}{\partial y_{\dot{\alpha}}\partial\tilde{y}_{\dot{\beta}}} = 1\,, \tag{3.13}$$

i.e., the metric determinant is 1. This condition is known as Plebanski's first heavenly equation [2].[1] It follows from equating the last of (3.9) with its definition $\Sigma^{22} = \theta^{2\dot{\alpha}} \wedge \theta^{2}{}_{\dot{\alpha}}$ in this frame.

---

[1] A broader treatment including the second heavenly equation as well as the more general case of quaternionic geometry can be found in [3–5].

The frame $\mathbf{e}_{\alpha\dot\alpha}$ of the tangent bundle can be taken to be

$$\mathbf{e}_{1\dot\alpha} = \frac{\partial}{\partial y^{\dot\alpha}}\,, \qquad \mathbf{e}_{2\dot\alpha} = \Omega_{\dot\alpha}{}^{\dot\beta}\,\frac{\partial}{\partial \tilde{y}^{\dot\beta}}\,. \tag{3.14}$$

It is dual to $\boldsymbol{\theta}^{\alpha\dot\alpha}$ when $\mathbf{e}_{2\dot\beta} \lrcorner\, \boldsymbol{\theta}^{2\dot\alpha} = \Omega_{\dot\beta\dot\gamma}\,\Omega^{\dot\alpha\dot\gamma} \stackrel{!}{=} \delta^{\dot\alpha}_{\dot\beta}$, i.e., when (3.13) is satisfied. The heavenly equation arises as the integrability condition $[\bar{L}_{\dot\alpha}, \bar{L}_{\dot\beta}] = 0$ of the Lax pair

$$\bar{L}_{\dot\alpha} := \lambda^\alpha\,\mathbf{e}_{\alpha\dot\alpha} = \lambda_2\,\frac{\partial}{\partial y^{\dot\alpha}} - \lambda_1\,\Omega_{\dot\alpha}{}^{\dot\beta}\frac{\partial}{\partial\tilde{y}^{\dot\beta}}\,. \tag{3.15}$$

Hence, it is a classical integrable system. Since they form an involutive distribution, the $\bar{L}_{\dot\alpha}$ provide (0, 1)-vectors of a new complex structure on M for every value of the spectral parameter $\lambda_\alpha \in \mathbb{P}^1$. In total, one finds that there are not just three but a continuous family of complex structures on M parametrized by points on $\mathbb{P}^1$. This provides a natural segway into the twistor correspondence for hyperkähler manifolds.

3.2 The Non-linear Graviton

Twistor space from space-time. Let (M, g) be a given self-dual space-time. Just as for flat space, the twistor space of M is given by the projectivized 2-spinor bundle of M. Locally, it has the structure of a product $\mathbb{P}\mathscr{T} \simeq M \times \mathbb{P}^1$. It is made into an almost complex manifold by taking the $(1, 0)$-forms to be the span of

$$\tau = \lambda^\alpha(\mathrm{d}\lambda_\alpha + \Gamma_\alpha{}^\beta\lambda_\beta)\,, \qquad \lambda_\alpha\,\boldsymbol{\theta}^{\alpha\dot\alpha}\,. \tag{3.16}$$

This provides the so-called Atiyah-Hitchin-Singer almost complex structure on $\mathbb{P}\mathscr{T}$ for vacuum as well as non-vacuum self-dual space-times. It becomes an integrable almost complex structure if these $(1, 0)$-forms span a differential ideal of the exterior algebra over $M \times \mathbb{P}^1$. It is a classic calculation in twistor theory that this happens if and only if $\Psi_{\alpha\beta\gamma\delta} = 0$ (see [6] for a review). That is, $\mathbb{P}\mathscr{T}$ is a complex manifold if and only if M is self-dual [7, 8].

We will focus on the SD vacuum case. A short computation shows that[2]

$$\tau \wedge \mathrm{d}\tau = -\tau \wedge \lambda^\alpha\,\lambda^\beta\,R_{\alpha\beta} = 0\,. \tag{3.17}$$

A holomorphic analogue of Frobenius' theorem then implies that the $(1, 0)$-vector fields in the kernel of τ form an integrable distribution. This endows $\mathbb{P}\mathscr{T}$ with a holomorphic fibration $\mathbb{P}\mathscr{T} \to \mathbb{P}^1$, with the fibers being the integral surfaces of

[2] It is easiest to find this by treating τ as a 1-form on non-projective twistor space $\mathscr{T} = M \times \mathbb{C}^2$ and computing its standard exterior derivative.

this distribution. Taking the flat ASD spin connection to be $\Gamma_{\alpha\beta} = 0$, we see from $\mathrm{d}\Sigma^{\alpha\beta} = 0$ that the fiber over every point $\lambda_\alpha \in \mathbb{P}^1$ comes equipped with a holomorphic symplectic structure

$$\Sigma = \lambda_\alpha \lambda_\beta \Sigma^{\alpha\beta} = \lambda_\alpha \theta^{\alpha\dot\alpha} \wedge \lambda_\beta \theta^\beta{}_{\dot\alpha} \,. \tag{3.18}$$

Indeed, $\mathrm{d}\Sigma = 0$ modulo terms containing $\mathrm{d}\lambda_\alpha$ which drop at fixed $\lambda_\alpha \in \mathbb{P}^1$.

Space-time from twistor space. Conversely, twistor theory allows one to construct SD vacuum space-times from 'curved' twistor spaces $\mathbb{P}\mathscr{T}$. A large class of curved twistor spaces can be explicitly obtained by deforming the complex structure of 'flat' twistor space $\mathbb{PT} = \mathscr{O}(1) \oplus \mathscr{O}(1) \to \mathbb{P}^1$. With such a deformation in hand, the key result we use for the reconstruction of space-time is Penrose's non-linear graviton theorem:

Theorem 3.1 (Penrose [9, 10]) *There is a one-to-one correspondence between:*

- *suitably convex regions of SD vacuum 4-manifolds* (M, g), *and*
- *Complex 3-folds* $\mathbb{P}\mathscr{T}$ *that are complex deformations of a neighbourhood of a projective line in* $\mathbb{PT}$ *and admit*

1. *a holomorphic fibration* $\mathbb{P}\mathscr{T} \to \mathbb{P}^1$,
2. *a 4-complex parameter family of sections each with normal bundle* $\mathscr{O}(1) \oplus \mathscr{O}(1)$,
3. *an* $\mathscr{O}(2)$*-valued holomorphic symplectic form* Σ *on each fiber, and*
4. *an anti-holomorphic involution* $j : \mathbb{P}\mathscr{T} \to \mathbb{P}\mathscr{T}$ *that induces the antipodal map on* $\mathbb{P}^1$ *and picks a 4-real parameter family of sections invariant under it.*

Here, $\mathscr{O}(n) \to \mathbb{P}\mathscr{T}$ is defined by pullback along the fibration $\mathbb{P}\mathscr{T} \to \mathbb{P}^1$.

We continue to use both non-holomorphic coordinates $(x^{\alpha\dot\alpha}, \lambda_\alpha)$ as well as holomorphic coordinates $(\mu^{\dot\alpha} = x^{\alpha\dot\alpha}\lambda_\alpha, \lambda_\alpha)$ on flat twistor space $\mathbb{PT}$. **The coordinates on space-time M will henceforth be denoted $y^{\alpha\dot\alpha}$ to avoid confusion**. The almost complex structure of $\mathbb{P}\mathscr{T}$ can be represented by expressing its $(1, 0)$-forms as deformations of those of $\mathbb{PT}$:

$$\begin{aligned} \theta^0 &= \mathrm{D}\lambda - V^0 \,, \qquad & V^0 &\in \Omega^{0,1}(\mathbb{PT}, \mathscr{O}(2)) \,, \\ \theta^{\dot\alpha} &= e^{\dot\alpha} - V^{\dot\alpha} \,, \qquad & V^{\dot\alpha} &\in \Omega^{0,1}(\mathbb{PT}, \mathscr{O}(1)) \,, \end{aligned} \tag{3.19}$$

where $\mathrm{D}\lambda \equiv \langle\lambda\, \mathrm{d}\lambda\rangle$ and $e^{\dot\alpha} = \lambda_\alpha\, \mathrm{d}x^{\alpha\dot\alpha}$ were introduced in (2.15). In analogy with 2d CFT, the vector field $V \equiv V^0\, \partial_0 + V^{\dot\alpha}\, \partial_{\dot\alpha} \in \Omega^{0,1}(\mathbb{PT}, T^{1,0}\mathbb{PT})$ will be called a Beltrami differential.

Remembering the twisted exterior derivative (2.18) of $\mathscr{O}(n)$-valued forms on $\mathbb{PT}$, a little work allows us to determine the structure equations associated to $\theta^0, \theta^{\dot\alpha}$:

$$\begin{aligned} \mathrm{d}\theta^0 &= -\theta^0 \wedge \mathcal{L}_{\partial_0} V^0 - \theta^{\dot\beta} \wedge \mathcal{L}_{\partial_{\dot\beta}} V^0 - N^0 \,, \\ \mathrm{d}\theta^{\dot\alpha} &= \theta^0 \wedge (\bar e^{\dot\alpha} - \mathcal{L}_{\partial_0} V^{\dot\alpha}) - \theta^{\dot\beta} \wedge \mathcal{L}_{\partial_{\dot\beta}} V^{\dot\alpha} - N^{\dot\alpha} \,, \end{aligned} \tag{3.20}$$

where

$$N^0 = \bar\partial V^0 + V^0 \wedge \mathcal{L}_{\partial_0} V^0 + V^{\dot\beta} \wedge \mathcal{L}_{\partial_{\dot\beta}} V^0 ,$$
$$N^{\dot\alpha} = \bar\partial V^{\dot\alpha} + V^0 \wedge (\mathcal{L}_{\partial_0} V^{\dot\alpha} - \bar{e}^{\dot\alpha}) + V^{\dot\beta} \wedge \mathcal{L}_{\partial_{\dot\beta}} V^{\dot\alpha} \tag{3.21}$$

are components of the Nijenhuis tensor $N = N^0\,\partial_0 + N^{\dot\alpha}\,\partial_{\dot\alpha}$.[3] In these expressions, $\bar\partial$ is the Dolbeault operator on $\mathbb{PT}$ defined in (2.19). The almost complex structure is integrable if and only if $N = 0$, for then the deformed $(1, 0)$-forms (3.19) span a differential ideal.

Following Theorem 3.1, we assume $N = 0$. The holomorphic fibration $\mathbb{P}\mathscr{T} \to \mathbb{P}^1$ can be modeled upon the fibration $\mathbb{PT} \to \mathbb{P}^1$ by setting $V^0 = 0$. In this case, θ^0 reduces to $\mathrm{D}\lambda$ and the flat twistor coordinate λ_α continues to be a holomorphic coordinate along the base of $\mathbb{P}\mathscr{T} \to \mathbb{P}^1$. On each fiber, the required symplectic $(2, 0)$-form can be taken to be

$$\Sigma = \theta^{\dot\alpha} \wedge \theta_{\dot\alpha} . \tag{3.22}$$

Computing its exterior derivative using (3.20) yields

$$\mathrm{d}\Sigma = 2\big[\theta^0 \wedge (\bar{e}^{\dot\alpha} - \mathcal{L}_{\partial_0} V^{\dot\alpha}) \wedge \theta_{\dot\alpha} + \theta_{\dot\alpha} \wedge \theta_{\dot\beta} \wedge \mathcal{L}_{\partial^{\dot\beta}} V^{\dot\alpha} - N^{\dot\alpha} \wedge \theta_{\dot\alpha}\big] . \tag{3.23}$$

Setting $V^0 = 0$ and $N^{\dot\alpha} = 0$, we see that $\mathrm{d}\Sigma = 0$ modulo terms containing $\mathrm{D}\lambda$ if and only if

$$\mathcal{L}_{\partial^{\dot\beta}} V^{\dot\alpha} - \mathcal{L}_{\partial^{\dot\alpha}} V^{\dot\beta} = 0 . \tag{3.24}$$

This is equivalent to saying that V is vertically divergence free: $\mathcal{L}_{\partial_{\dot\alpha}} V^{\dot\alpha} = 0$.

Motivated by this, we introduce a holomorphic Poisson bracket on the fibers of $\mathbb{PT} \to \mathbb{P}^1$:

$$\{\omega, \eta\} := \epsilon^{\dot\alpha\dot\beta}\, \mathcal{L}_{\partial_{\dot\alpha}}\omega \wedge \mathcal{L}_{\partial_{\dot\beta}}\eta \tag{3.25}$$

on $(0, \bullet)$-forms ω, η on $\mathbb{PT}$. We can locally solve the constraints $V^0 = 0$ and (3.24) in terms of a hamiltonian $h \in \Omega^{0,1}(\mathbb{PT}, \mathcal{O}(2))$ by setting

$$V = \{h, \cdot\} = \epsilon^{\dot\alpha\dot\beta}\, \mathcal{L}_{\partial_{\dot\alpha}} h\, \partial_{\dot\beta} , \tag{3.26}$$

with h now encoding the data of the complex structure deformation. It is defined up to gauge transformations generated by Poisson diffeomorphisms. For instance, linearized gauge transformations take the form

$$\delta h = \bar\partial\chi + \{h, \chi\} , \qquad \chi \in \Omega^0(\mathbb{PT}, \mathcal{O}(2)) . \tag{3.27}$$

[3] Strictly speaking, this is a perturbative analogue of the actual Nijenhuis tensor that nevertheless captures all its non-linearities.

Using (3.26), the Nijenhuis tensor can be expressed in terms of a "torsion" T (unrelated to space-time torsion):

$$N = \{T, \cdot\}, \qquad T := \bar{\partial} h + \frac{1}{2}\{h, h\}. \tag{3.28}$$

In what follows, we will demand $T = 0$ as our criterion of integrability. Although this is mildly stronger than the minimal requirement $\mathcal{L}_{\partial_{\dot{\alpha}}} T = 0$, it will be needed to generate solutions of the heavenly Eq. (3.13) from twistor sigma models in the next section.

To construct an SD vacuum space-time M from this data, we begin by constructing holomorphic rational curves in $\mathbb{P}\mathscr{T}$. Let $Y \simeq \mathbb{P}^1$ be such a curve. It is given by rational maps $(\mu^{\dot{\alpha}}(\sigma), \lambda_\alpha(\sigma)) : \mathbb{P}^1 \to \mathbb{P}\mathscr{T}$, where $\sigma_\alpha \in \mathbb{P}^1$ are homogeneous coordinates parametrizing Y. For Y to embed holomorphically in $\mathbb{P}\mathscr{T}$, it needs to satisfy the defining equation of a (pseudo-)holomorphic curve in a (almost) complex manifold [11]

$$T^{0,1}\mathbb{P}\mathscr{T}|_Y \simeq T^{0,1}\mathbb{P}^1. \tag{3.29}$$

Concretely, this says that the standard (0, 1)-vector field $\langle\hat{\sigma}\,\sigma\rangle\,\sigma \cdot \partial_{\hat{\sigma}}$ on $\mathbb{P}^1$ must be in the kernel of the (1, 0)-forms on $\mathbb{P}\mathscr{T}$, resulting in the PDEs

$$\langle\hat{\sigma}\,\sigma\rangle\,\sigma_\alpha\,\frac{\partial}{\partial\hat{\sigma}_\alpha} \lrcorner\,(\langle\lambda\,\mathrm{d}\lambda\rangle|_Y) = 0\,, \qquad \langle\hat{\sigma}\,\sigma\rangle\,\sigma_\alpha\,\frac{\partial}{\partial\hat{\sigma}_\alpha} \lrcorner\,(\theta^{\dot{\alpha}}|_Y) = 0\,. \tag{3.30}$$

Up to GL(2, $\mathbb{C}$) rotations, the first of these is solved by $\lambda_\alpha(\sigma) = \sigma_\alpha$, whence we can use λ_α itself as the curve parameter.

Thus, Y gives a global holomorphic section of the fibration $\mathbb{P}\mathscr{T} \to \mathbb{P}^1$,

$$(\mu^{\dot{\alpha}} = \mu^{\dot{\alpha}}(\lambda),\, \lambda_\alpha) : \mathbb{P}^1 \to \mathbb{P}\mathscr{T}, \tag{3.31}$$

where $\mu^{\dot{\alpha}}(\lambda)$ is a rational function homogeneous of degree 1 in λ_α. By assumption, each section of $\mathbb{P}\mathscr{T} \to \mathbb{P}^1$ has normal bundle $\mathscr{O}(1) \oplus \mathscr{O}(1)$. Since $H^0(\mathbb{P}^1, \mathscr{O}(1) \oplus \mathscr{O}(1)) = \mathbb{C}^4$, Kodaira theory allows us to construct a complex space-time $M_{\mathbb{C}}$ as the 4-complex-dimensional moduli space of such sections. For every $y \in M_{\mathbb{C}}$, the curve is given by a degree 1 rational function $F^{\dot{\alpha}}(y, \lambda)$,

$$Y\,:\quad \mu^{\dot{\alpha}}(\lambda) = F^{\dot{\alpha}}(y, \lambda)\,. \tag{3.32}$$

It must satisfy the remaining equation in (3.30) but with $\langle\hat{\sigma}\,\sigma\rangle\,\sigma \cdot \partial_{\hat{\sigma}}$ replaced by our old friend $\bar{\partial}_0 = \langle\hat{\lambda}\,\lambda\rangle\,\lambda \cdot \partial_{\hat{\lambda}}$,

$$\bar{\partial}_0 \lrcorner \left(\theta^{\dot{\alpha}}|_Y\right) = 0 \implies \bar{\partial}_0 F^{\dot{\alpha}} + \bar{\partial}_0 \lrcorner\,(\mathcal{L}_{\partial^{\dot{\alpha}}} h|_Y) = 0\,. \tag{3.33}$$

Remember that in this PDE, $\bar{\partial}_0$ is evaluated at fixed y (instead of fixed x).

Lastly, we can obtain a real hyperkähler slice $M \subset M_{\mathbb{C}}$ as the space of those sections that are left invariant by the involution j mentioned in the last point of Theorem 3.1. If Z^A and $j(Z^A)$ both lie on a curve Y, then they uniquely determine the curve and it corresponds to a real point $y \in M$. For example, for $\mathbb{R}^4$, such an involution was given by quaternionic conjugation $j(Z^A) = \hat{Z}^A$ in (2.9). In practice, we will never need to actually find such a j, because our calculations will always be analytic in $y \in M_{\mathbb{C}}$. As such, the final results of most of our calculations will be signature-agnostic.

Reconstruction of the hyperkähler metric on M. The map

$$p : M \times \mathbb{P}^1 \to \mathbb{P}\mathscr{T}\,, \qquad (y, \lambda) \mapsto Z^A = (F^{\dot\alpha}, \lambda_\alpha) \tag{3.34}$$

is a diffeomorphism that realizes $\mathbb{P}\mathscr{T}$ as the spinor bundle of $M \subset M_{\mathbb{C}}$. Since $\bar\partial_0$ is in the kernel of the 1-forms on $M \times \mathbb{P}^1$ that point along M, we can trivially rewrite (3.33) as

$$\bar\partial_0 \lrcorner\, p^*\theta^{\dot\alpha} = 0 \implies \bar\partial_0 F^{\dot\alpha} + \bar\partial_0 \lrcorner\, p^* \mathcal{L}_{\partial^{\dot\alpha}} h = 0\,. \tag{3.35}$$

Using this, we can show that the 2-form $\Sigma = \theta^{\dot\alpha} \wedge \theta_{\dot\alpha}$ satisfies

$$\begin{aligned} \mathcal{L}_{\bar\partial_0} p^*\Sigma &= \bar\partial_0 \lrcorner\, p^*\mathrm{d}\Sigma + \mathrm{d}(\bar\partial_0 \lrcorner\, p^*\Sigma) \\ &= 0 \qquad \mathrm{mod}\ \mathrm{D}\lambda\,, \end{aligned} \tag{3.36}$$

where the first term vanished because $\mathrm{d}\Sigma = 0$ modulo terms containing $\mathrm{D}\lambda$ (i.e., Σ is closed at constant λ_α) and the second term vanished by virtue of (3.35).

Thus, Σ is a holomorphic 2-form of homogeneity 2 in λ_α. By Liouville's theorem applied to every curve $Y \simeq \mathbb{P}^1$, there exists a triplet of 2-forms $\Sigma^{\alpha\beta} = \Sigma^{(\alpha\beta)}$ on M such that

$$p^*\Sigma = \lambda_\alpha\, \lambda_\beta\, \Sigma^{\alpha\beta}(y) \qquad \mathrm{mod}\ \mathrm{D}\lambda\,. \tag{3.37}$$

It follows by construction that the 2-forms $\Sigma^{\alpha\beta}$ obey $\mathrm{d}\Sigma^{\alpha\beta} = 0$. Since $\Sigma \wedge \Sigma = 0$, they are constrained by $\Sigma^{(\alpha\beta} \wedge \Sigma^{\gamma\delta)} = 0$. This is known as a 'simplicity' constraint. It implies the existence of a frame $\boldsymbol{\theta}^{\alpha\dot\alpha}$ on M for which [12]:

$$\Sigma^{\alpha\beta} = \boldsymbol{\theta}^{\alpha\dot\alpha} \wedge \boldsymbol{\theta}^{\beta}{}_{\dot\alpha}\,. \tag{3.38}$$

With this frame, the hyperkähler metric on M is recovered to be $\mathrm{d}s^2 = \boldsymbol{\theta}^{\alpha\dot\alpha}\, \boldsymbol{\theta}_{\alpha\dot\alpha}$; the holonomy reduction to $\mathrm{SU}(2)$ follows as a consequence of $\mathrm{d}\Sigma^{\alpha\beta} = 0$.

Modulo $\mathrm{D}\lambda$, it follows from (3.37) and (3.38) that $p^*\theta^{\dot\alpha} \wedge p^*\theta_{\dot\alpha} = \lambda_\alpha\, \boldsymbol{\theta}^{\alpha\dot\alpha} \wedge \lambda_\beta\, \boldsymbol{\theta}^{\beta}{}_{\dot\alpha}$. This determines $\theta^{\dot\alpha}$ in terms of space-time data up to a rotation $H^{\dot\alpha}{}_{\dot\beta}(y, \lambda, \hat\lambda) \in \mathrm{SL}(2, \mathbb{C})$ of the dotted indices [13],

$$p^*\theta^{\dot\alpha} = \lambda_\beta\, H^{\dot\alpha}{}_{\dot\beta}\, \boldsymbol{\theta}^{\beta\dot\beta} \qquad \mathrm{mod}\ \mathrm{D}\lambda\,. \tag{3.39}$$

For each $y \in M$, the matrix $H^{\dot\alpha}{}_{\dot\beta}$ has homogeneity 0 in $\lambda_\alpha, \hat\lambda_\alpha$ and can be interpreted as a spin frame on the bundle of dotted spinors $\mathcal{O}(0) \oplus \mathcal{O}(0) \to Y$ over the curve Y. Its unimodularity can be reexpressed as

$$\epsilon_{\dot\alpha\dot\beta}\, H^{\dot\alpha}{}_{\dot\gamma}\, H^{\dot\beta}{}_{\dot\delta} = \epsilon_{\dot\gamma\dot\delta}\,, \tag{3.40}$$

which ensures that it drops out of $p^*\theta^{\dot\alpha} \wedge p^*\theta_{\dot\alpha}$.

Further acting with $\mathcal{L}_{\bar\partial_0}$ on (3.39), we can find a PDE for the spin frame:

$$\begin{aligned}\lambda_\beta\, \bar\partial_0 H^{\dot\alpha}{}_{\dot\beta}\, \boldsymbol{\theta}^{\beta\dot\beta} &= \mathcal{L}_{\bar\partial_0} p^*\theta^{\dot\alpha} = \bar\partial_0 \lrcorner\, p^*\mathrm{d}\theta^{\dot\alpha} = -\big(\bar\partial_0 \lrcorner\, p^*\mathcal{L}_{\partial_{\dot\beta}}\mathcal{L}_{\partial^{\dot\alpha}} h\big)\, p^*\theta^{\dot\beta} \qquad \operatorname{mod} \mathrm{D}\lambda \\ \Longrightarrow\ \bar\partial_0 H^{\dot\alpha}{}_{\dot\beta} &+ \bar\partial_0 \lrcorner\, p^*\mathcal{L}_{\partial^{\dot\alpha}}\mathcal{L}_{\partial_{\dot\gamma}} h\ H^{\dot\gamma}{}_{\dot\beta} = 0\,,\end{aligned} \tag{3.41}$$

having applied the second structure equation in (3.20) with data $V^0 = 0$, $V^{\dot\alpha} = -\mathcal{L}_{\partial^{\dot\alpha}} h$ and $N = 0$. We remark that in deriving this, $N = 0$ can be replaced by the weaker condition $\bar\partial_0 \lrcorner\, p^* N^{\dot\alpha} = 0$ [14], though we will ignore this subtlety for the purposes of this thesis.

3.3 Twistor Sigma Models

We can use the 2-forms $\Sigma^{\alpha\beta}$ generated by the non-linear graviton construction to produce explicit hyperkähler metrics of the form (3.11) on M. Much like the case of linear fields in Chap. 2, this will entail finding explicit integral formulae for solutions Ω to Plebanski's heavenly Eq. (3.13), leading us to the namesake sigma models of this thesis.

Curves and boundary conditions. Let us work on a complexified coordinate patch $y^{\alpha\dot\alpha} \in M_{\mathbb{C}}$. Our calculations in this section will be analytic in $y^{\alpha\dot\alpha}$, but it is useful to keep a Euclidean real slice $M \subset M_{\mathbb{C}}$ at the back of one's mind. Let $\kappa_{1\alpha}, \kappa_{2\alpha}$ be a basis of undotted spinors, normalized with the convention $\langle 1\, 2\rangle \equiv \langle \kappa_1\, \kappa_2\rangle = -1$. In this basis, we set

$$\lambda_\alpha = \lambda_1\, \kappa_{1\alpha} + \lambda_2\, \kappa_{2\alpha}\,, \qquad y^{\dot\alpha} = y^{\alpha\dot\alpha}\kappa_{1\alpha}\,, \qquad \tilde y^{\dot\alpha} = y^{\alpha\dot\alpha}\kappa_{2\alpha}\,. \tag{3.42}$$

Using a general spin frame κ_1, κ_2 will prove extremely helpful in the study of amplitudes in the next chapter, where the basis spinors will get identified with spinor-helicity variables of negative helicity gravitons. Later on, factors of $\langle 1\, 2\rangle$ will also be reinstated by invoking standard scaling properties of amplitudes.

In the previous section, we introduced holomorphic curves (3.32) in deformed twistor spaces $\mathbb{P}\mathcal{T}$ by realizing them as global sections of the holomorphic fibration $\mathbb{P}\mathcal{T} \to \mathbb{P}^1$. The coordinates $y^{\dot\alpha}$, $\tilde y^{\dot\alpha}$ acted as moduli for their associated curve Y. Henceforth, the fundamental question that will inspire much of our investigation is the following: *can we find a sigma model governing the holomorphic embeddings*

$Y \simeq \mathbb{P}^1 \hookrightarrow \mathbb{P}\mathcal{T}$? That is, can we find a field theory living on Y whose equations of motion give the PDE (3.33)?

To study this question, first decompose the curved incidence relations as

$$\mu^{\dot\alpha} = F^{\dot\alpha}(y,\lambda) = \lambda_1\, y^{\dot\alpha} + \lambda_2\, \tilde y^{\dot\alpha} + m^{\dot\alpha}(y,\lambda) \tag{3.43}$$

where $y^{\alpha\dot\alpha}\lambda_\alpha = \lambda_1\, y^{\dot\alpha} + \lambda_2\, \tilde y^{\dot\alpha}$ acts as a 'zero-mode', and $m^{\dot\alpha}(y,\lambda)$ encodes perturbations generated by turning on the complex structure deformation h. Substituting (3.43) into (3.33) yields the PDE satisfied by the perturbations,

$$\bar\partial_0 m^{\dot\alpha} + \bar\partial_0 \lrcorner (\mathcal{L}_{\partial^{\dot\alpha}} h|_Y) = 0\,. \tag{3.44}$$

Using the Cauchy kernel of the Dolbeault operator of $\mathbb{P}^1$ acting on sections of $\mathcal{O}(1)$, we can turn this into an integral equation:

$$m^{\dot\alpha}(y,\lambda) = -\frac{1}{2\pi \mathrm{i}}\int_{Y'} \frac{\mathrm{D}\lambda'}{\langle\lambda\,\lambda'\rangle}\frac{\langle 1\,\lambda\rangle}{\langle 1\,\lambda'\rangle}\frac{\langle 2\,\lambda\rangle}{\langle 2\,\lambda'\rangle} \wedge \mathcal{L}_{\partial^{\dot\alpha}} h|_{Y'}\,, \tag{3.45}$$

where Y' denotes the same curve as Y but with curve coordinates λ'_α, and $\langle i\,\lambda\rangle \equiv \langle \kappa_i\,\lambda\rangle$, etc. As with any Green's function, the Cauchy kernel requires a set of boundary conditions that are here represented by a pair of 'reference spinors' κ_1, κ_2 at which $m^{\dot\alpha}$ is deemed to vanish. With hindsight, we have identified these with our choice of spin frame.

With these boundary conditions, we can factorize $m^{\dot\alpha} = \lambda_1\lambda_2\, \mathrm{m}^{\dot\alpha}$, with $\mathrm{m}^{\dot\alpha}(y,\lambda)$ being $\mathcal{O}(-1)$-valued rational functions that are regular at $\lambda = \kappa_1, \kappa_2$. They are determined by the integral equation

$$\mathrm{m}^{\dot\alpha}(y,\lambda) = \frac{m^{\dot\alpha}(y,\lambda)}{\lambda_1\lambda_2} = -\frac{1}{2\pi\mathrm{i}}\int_{Y'} \frac{\mathrm{D}\lambda'}{\langle\lambda\,\lambda'\rangle} \wedge \frac{\mathcal{L}_{\partial^{\dot\alpha}} h|_{Y'}}{\lambda'_1\lambda'_2} \tag{3.46}$$

obtained from dividing (3.45) by $\lambda_1\lambda_2 = -\langle 1\,\lambda\rangle\,\langle 2\,\lambda\rangle$. Thus, our curves take the form

$$\mu^{\dot\alpha} = F^{\dot\alpha}(y,\lambda) = \lambda_1\, y^{\dot\alpha} + \lambda_2\, \tilde y^{\dot\alpha} + \lambda_1\lambda_2\, \mathrm{m}^{\dot\alpha}(y,\lambda)\,, \tag{3.47}$$

and satisfy the boundary conditions

$$F^{\dot\alpha}(y,\kappa_1) = y^{\dot\alpha}\,, \qquad F^{\dot\alpha}(y,\kappa_2) = \tilde y^{\dot\alpha}\,. \tag{3.48}$$

These fix the moduli of the curve so that $\mathrm{m}^{\dot\alpha}(y,\lambda)$ contain no newer moduli.

Sigma model and equations of motion. We now construct a second order action governing the perturbations $m^{\dot\alpha}$.[4] The fields of our theory will be the complex symplectic

---

[4] This is distinct from twistor strings, where one usually constructs a first order sigma model governing the full map $F^{\dot\alpha}(y,\lambda)$ but requires the introduction of dual twistor fields.

bosons $m^{\dot\alpha}$ and their quaternionic conjugates $\hat{m}^{\dot\alpha}$. In terms of these, we introduce a *twistor sigma model* as the field theory given by the action

$$S_h[m](y) = \frac{1}{2\pi \mathrm{i}} \int_Y \frac{\mathrm{D}\lambda}{\lambda_1^2 \lambda_2^2} \wedge \left(\frac{1}{2} m^{\dot\alpha}\, \bar\partial|_Y m_{\dot\alpha} + h|_Y \right) , \tag{3.49}$$

where $\bar\partial|_Y = \bar{e}^0\, \bar\partial_0$ denotes the standard anti-holomorphic exterior derivative on $\mathbb{P}^1$. This action is a functional of the fields $m^{\dot\alpha}$ as well as an ordinary function of the curve moduli y. It is coupled to the complex structure deformation h simply by pullback to Y.

Notable is the presence of poles in the integration measure $\mathrm{D}\lambda/\lambda_1^2\lambda_2^2$, which act as defects coupled to our sigma models. These are necessitated for reasons of homogeneity. The ingredients $\mathrm{D}\lambda$, $m^{\dot\alpha}\, \bar\partial|_Y m_{\dot\alpha}$ and h all have weight 2 in λ_α, so we need to introduce poles in the measure to cancel the weight and give a genuine integrand on $\mathbb{P}^1$. The natural choice of double poles at $\lambda_1 = 0$ and $\lambda_2 = 0$ arises from our boundary conditions on the fields. Since $m^{\dot\alpha} = \lambda_1 \lambda_2\, \mathrm{m}^{\dot\alpha}$, the kinetic term secretly has no poles and the $\mathbb{P}^1$-integral can converge. By the same logic, the interaction term containing $h|_Y$ can only be convergent in a gauge where h vanishes to second order at both poles, i.e.,

$$h = \lambda_1^2\, \lambda_2^2\, \omega\,, \qquad \omega \in \Omega^{0,1}(\mathbb{PT}, \mathscr{O}(-2))\,. \tag{3.50}$$

Recalling the linear Penrose transform discussed in Sect. 2.3, this indicates that h contains the degree of freedom ω of a single scalar field on space-time. We are on the right track to find the Kähler scalar!

Next, we compute the equations of motion of (3.49). To do this, set $h = h_0\, \bar{e}^0 + h_{\dot\alpha}\, \bar{e}^{\dot\alpha}$. Notice that our model depends on h only through $\mathrm{D}\lambda \wedge h|_Y = \mathrm{D}\lambda \wedge \bar{e}^0\, \bar\partial_0 \lrcorner\, (h|_Y)$. Using the incidence relations map

$$p\ :\quad (y, \lambda) \mapsto \left(x^{\alpha\dot\alpha} = \frac{\hat\lambda^\alpha F^{\dot\alpha}(y,\lambda) - \lambda^\alpha \hat{F}^{\dot\alpha}(y,\lambda)}{\langle \hat\lambda\, \lambda \rangle}, \lambda \right) \tag{3.51}$$

we find

$$\bar\partial_0 \lrcorner\, (h|_Y) = h_0|_Y - h_{\dot\alpha}|_Y\, \frac{\hat\lambda_\alpha\, \bar\partial_0|_y x^{\alpha\dot\alpha}}{\langle \hat\lambda\, \lambda \rangle} = h_0|_Y + h_{\dot\alpha}|_Y \left(F^{\dot\alpha} - \frac{\bar\partial_0 \hat{F}^{\dot\alpha}}{\langle \hat\lambda\, \lambda \rangle} \right), \tag{3.52}$$

where we have clarified that $\bar\partial_0|_y$ denotes $\bar\partial_0$ evaluated at fixed y. Since $\delta F^{\dot\alpha} = \delta m^{\dot\alpha}$ and $\delta \hat{F}^{\dot\alpha} = \delta \hat{m}^{\dot\alpha}$ under variations of $m^{\dot\alpha}$, $\hat{m}^{\dot\alpha}$, varying the action (3.49) with respect to $m^{\dot\alpha}$ gives

$$\delta_m S_h = \frac{1}{2\pi\mathrm{i}} \int_Y \frac{\mathrm{D}\lambda \wedge \bar{e}^0}{\lambda_1^2 \lambda_2^2}\, \delta m^{\dot{\alpha}} \left[\bar{\partial}_0 m_{\dot{\alpha}} + (\partial_{\dot{\alpha}} h_0 + h_{\dot{\alpha}})|_Y + \partial_{\dot{\alpha}} h_{\dot{\beta}}|_Y \left(F^{\dot{\beta}} - \frac{\bar{\partial}_0 \hat{F}^{\dot{\beta}}}{\langle \hat{\lambda}\, \lambda \rangle} \right) \right] + \frac{1}{2\pi\mathrm{i}} \int_Y \frac{\mathrm{D}\lambda \wedge \bar{e}^0}{\lambda_1^2 \lambda_2^2}\, \bar{\partial}_0 (m^{\dot{\alpha}} \delta m_{\dot{\alpha}})\,. \tag{3.53}$$

In the last term, we can use the boundary conditions to substitute $m^{\dot{\alpha}} = \lambda_1 \lambda_2\, \mathfrak{m}^{\dot{\alpha}}$ with $\mathfrak{m}^{\dot{\alpha}}$ regular at the two poles. It then vanishes by Stokes' theorem:

$$\frac{1}{2\pi\mathrm{i}} \int_Y \frac{\mathrm{D}\lambda \wedge \bar{e}^0}{\lambda_1^2 \lambda_2^2}\, \bar{\partial}_0 (m^{\dot{\alpha}} \delta m_{\dot{\alpha}}) = \frac{1}{2\pi\mathrm{i}} \int_Y \mathrm{D}\lambda \wedge \bar{e}^0\, \bar{\partial}_0 (\mathfrak{m}^{\dot{\alpha}} \delta \mathfrak{m}_{\dot{\alpha}}) = 0\,. \tag{3.54}$$

Recycling the calculation in (3.52) with h replaced by $\mathcal{L}_{\partial_{\dot{\alpha}}} h = (\partial_{\dot{\alpha}} h_0 + h_{\dot{\alpha}})\, \bar{e}^0 + \partial_{\dot{\alpha}} h_{\dot{\beta}}\, \bar{e}^{\dot{\beta}}$, we reduce the remaining variation to

$$\delta_m S_h = \frac{1}{2\pi\mathrm{i}} \int_Y \frac{\mathrm{D}\lambda \wedge \bar{e}^0}{\lambda_1^2 \lambda_2^2}\, \delta m^{\dot{\alpha}} \left[\bar{\partial}_0 m_{\dot{\alpha}} + \bar{\partial}_0 \lrcorner (\mathcal{L}_{\partial_{\dot{\alpha}}} h|_Y) \right]. \tag{3.55}$$

Demanding that this vanish shows that the equation of motion of $m^{\dot{\alpha}}$ is precisely (3.44), the PDE for holomorphic perturbations.

The equations of motion of the complex conjugates of $m^{\dot{\alpha}}$ can be similarly deduced by varying S_h with respect to $\hat{m}^{\dot{\alpha}}$ (but keeping $m^{\dot{\alpha}}$ formally fixed as is standard for actions involving complex scalars). In the first instance, this leads to

$$\delta_{\hat{m}} S_h = \frac{1}{2\pi\mathrm{i}} \int_Y \frac{\mathrm{D}\lambda \wedge \bar{e}^0}{\lambda_1^2 \lambda_2^2}\, \frac{\delta \hat{m}^{\dot{\alpha}}}{\langle \hat{\lambda}\, \lambda \rangle} \left[-\bar{\partial}_{\dot{\alpha}} h_0|_Y - \bar{\partial}_{\dot{\alpha}} h_{\dot{\beta}}|_Y \left(F^{\dot{\beta}} - \frac{\bar{\partial}_0 \hat{F}^{\dot{\beta}}}{\langle \hat{\lambda}\, \lambda \rangle} \right) + \bar{\partial}_0 (h_{\dot{\alpha}}|_Y) \right] - \frac{1}{2\pi\mathrm{i}} \int_Y \frac{\mathrm{D}\lambda \wedge \bar{e}^0}{\lambda_1^2 \lambda_2^2}\, \bar{\partial}_0 \left(\frac{h_{\dot{\alpha}}|_Y\, \delta \hat{m}^{\dot{\alpha}}}{\langle \hat{\lambda}\, \lambda \rangle} \right). \tag{3.56}$$

Using the boundary conditions (3.50) on h, we again see that the last term drops out due to Stokes' theorem. The remaining variation can be simplified using the chain rule $\bar{\partial}_0 (h_{\dot{\alpha}}|_Y) = \bar{\partial}_0 h_{\dot{\alpha}}|_Y + \partial h_{\dot{\alpha}} / \partial x^{\beta\dot{\beta}}|_Y\, \bar{\partial}_0|_y x^{\beta\dot{\beta}}$ to yield the equation of motion

$$\left(\bar{\partial}_0 h_{\dot{\alpha}} - \bar{\partial}_{\dot{\alpha}} h_0 \right)|_Y + \partial_{\dot{\beta}} h_{\dot{\alpha}}|_Y\, \bar{\partial}_0 m^{\dot{\beta}} + \left(\bar{\partial}_{\dot{\beta}} h_{\dot{\alpha}} - \bar{\partial}_{\dot{\alpha}} h_{\dot{\beta}} \right)|_Y \left(F^{\dot{\beta}} - \frac{\bar{\partial}_0 \hat{F}^{\dot{\beta}}}{\langle \hat{\lambda}\, \lambda \rangle} \right) = 0\,. \tag{3.57}$$

Substituting for $\bar{\partial}_0 m^{\dot{\beta}}$ using the $m^{\dot{\alpha}}$ equation of motion (3.44), this can be recast as

$$(\bar{\partial}_0 h_{\dot{\alpha}} - \bar{\partial}_{\dot{\alpha}} h_0 + \{h_0, h_{\dot{\alpha}}\} - h^{\dot{\beta}} \partial_{\dot{\beta}} h_{\dot{\alpha}})|_Y$$
$$+(\bar{\partial}_{\dot{\beta}} h_{\dot{\alpha}} - \bar{\partial}_{\dot{\alpha}} h_{\dot{\beta}} + \{h_{\dot{\beta}}, h_{\dot{\alpha}}\})|_Y \left(F^{\dot{\beta}} - \frac{\bar{\partial}_0 \hat{F}^{\dot{\beta}}}{\langle \hat{\lambda}\, \lambda \rangle} \right) = 0\,, \tag{3.58}$$

written in terms of the Poisson bracket (3.25). It gives a constraint on the ambient complex structure h, but how do we interpret it?

Recall the "torsion" $T = \bar{\partial} h + \frac{1}{2}\{h, h\}$ introduced in (3.28). Writing it in components,

$$T = T_{0\dot{\alpha}}\, \bar{e}^0 \wedge \bar{e}^{\dot{\alpha}} + \frac{1}{2}\, T_{\dot{\beta}\dot{\alpha}}\, \bar{e}^{\dot{\beta}} \wedge \bar{e}^{\dot{\alpha}}\,, \tag{3.59}$$

we can easily determine the explicit expressions

$$\begin{aligned} T_{0\dot{\alpha}} &= \bar{\partial}_0 h_{\dot{\alpha}} - \bar{\partial}_{\dot{\alpha}} h_0 + \{h_0, h_{\dot{\alpha}}\} - h^{\dot{\beta}} \partial_{\dot{\beta}} h_{\dot{\alpha}}\,, \\ T_{\dot{\beta}\dot{\alpha}} &= \bar{\partial}_{\dot{\beta}} h_{\dot{\alpha}} - \bar{\partial}_{\dot{\alpha}} h_{\dot{\beta}} + \{h_{\dot{\beta}}, h_{\dot{\alpha}}\}\,. \end{aligned} \tag{3.60}$$

It follows that (3.58) is encoding the vanishing of the projection of T along Y:

$$\bar{\partial}_0 \lrcorner\, p^* T = 0\,, \tag{3.61}$$

with $p : M \times \mathbb{P}^1 \to \mathbb{P}\mathscr{T}$ given in (3.51). At the end of the day, we want to work in the category of complex structure deformations satisfying $T = 0$. So we conclude that the $\hat{m}^{\dot{\alpha}}$ equation of motion does not give any further constraints on the curves beyond (3.44).

Thus, up to a choice of boundary conditions, our sigma model governs the required holomorphic curves in $\mathbb{P}\mathscr{T}$.

Solutions of the heavenly equation. The prime utility of our twistor sigma model is that its on-shell action gives rise to a Kähler scalar on space-time. This is captured by the following result which forms the backbone of the chapters to follow.

Theorem 3.2 *Let $\mathbb{P}\mathscr{T}$ be a curved twistor space described by a hamiltonian complex structure deformation $h \in \Omega^{0,1}(\mathbb{PT}, \mathcal{O}(2))$ satisfying $T = 0$ and vanishing to second order at κ_1, κ_2. In complex coordinates $y^{\dot{\alpha}} = y^{\alpha\dot{\alpha}}\kappa_{1\alpha}$, $\tilde{y}^{\dot{\alpha}} = y^{\alpha\dot{\alpha}}\kappa_{2\alpha}$, the Kähler scalar solving Plebanski's heavenly Eq. (3.13) on the associated hyperkähler 4-manifold M is given by*

$$\Omega(y) = \epsilon_{\dot{\alpha}\dot{\beta}}\, y^{\dot{\alpha}} \tilde{y}^{\dot{\beta}} - S_h[m](y)\big|_{\text{on-shell}}\,. \tag{3.62}$$

The proof of this theorem follows by direct computation of the Kähler form. This can be shown in two steps.

Lemma 3.1 *First derivatives of the twistor sigma model's on-shell action yield*

$$\frac{\partial}{\partial y^{\dot\alpha}} S_h[m](y)\big|_{\text{on-shell}} = -\mathfrak{m}_{\dot\alpha}(y,\kappa_1)\,, \qquad \frac{\partial}{\partial \tilde y^{\dot\alpha}} S_h[m](y)\big|_{\text{on-shell}} = \mathfrak{m}_{\dot\alpha}(y,\kappa_2)\,. \tag{3.63}$$

Proof Much of the computation in deriving these is a repeat of that involved in getting the equations of motion. To outline our methods, let's obtain the $y^{\dot\alpha}$-derivative. Even before assuming that the action is on-shell or that $T = 0$, one can massage it into[5]

$$\begin{aligned}\frac{\partial}{\partial y^{\dot\alpha}} S_h[m](y) = \frac{1}{2\pi \mathrm{i}}\int_Y \frac{\mathrm{D}\lambda\wedge\bar e^0}{\lambda_1^2\lambda_2^2}\bigg[&\frac{1}{2}\frac{\partial m^{\dot\beta}}{\partial y^{\dot\alpha}}\,\bar\partial_0 m_{\dot\beta} + \frac{1}{2}\,m^{\dot\beta}\,\bar\partial_0\frac{\partial m_{\dot\beta}}{\partial y^{\dot\alpha}} - \frac{\partial F^{\dot\beta}}{\partial y^{\dot\alpha}}\,\bar\partial_0 m_{\dot\beta}\\ &+\frac{1}{2}\frac{\partial}{\partial y^{\dot\alpha}}\lrcorner\,\bar\partial_0\lrcorner\,p^*(\Sigma+2\,T) + \bar\partial_0\Big(\frac{\partial}{\partial y^{\dot\alpha}}\lrcorner\,p^*h\Big)\bigg]\,,\end{aligned} \tag{3.64}$$

where recall that $\Sigma = \theta^{\dot\alpha}\wedge\theta_{\dot\alpha}$. As usual, we can drop the last term using Stokes' theorem and the boundary conditions (3.48), (3.50) on $F^{\dot\alpha}$ and h. Using integrability $T = 0$ as well as the $m^{\dot\alpha}$ equation of motion $\bar\partial_0\lrcorner\,p^*\theta^{\dot\alpha} = 0$ (see (3.35)), we can also set $\bar\partial_0\lrcorner\,p^*(\Sigma + 2\,T) = 0$.

This leaves us with just the first line of (3.64). Substituting $F^{\dot\alpha} = \lambda_1 y^{\dot\alpha} + \lambda_2\tilde y^{\dot\alpha} + m^{\dot\alpha}$ along with $m^{\dot\alpha} = \lambda_1\lambda_2\,\mathfrak{m}^{\dot\alpha}$ then yields

$$\begin{aligned}\frac{\partial}{\partial y^{\dot\alpha}} S_h[m](y)\big|_{\text{on-shell}} &= \int_Y \frac{\mathrm{D}\lambda\wedge\bar e^0}{2\pi\mathrm{i}}\left[\frac{1}{2}\frac{\partial \mathfrak{m}^{\dot\beta}}{\partial y^{\dot\alpha}}\,\bar\partial_0\mathfrak{m}_{\dot\beta} + \frac{1}{2}\,\mathfrak{m}^{\dot\beta}\,\bar\partial_0\frac{\partial\mathfrak{m}_{\dot\beta}}{\partial y^{\dot\alpha}} - \left(\frac{\delta^{\dot\beta}_{\dot\alpha}}{\lambda_2} + \frac{\partial\mathfrak{m}^{\dot\beta}}{\partial y^{\dot\alpha}}\right)\bar\partial_0\mathfrak{m}_{\dot\beta}\right]\\ &= \int_Y \frac{\mathrm{D}\lambda\wedge\bar e^0}{2\pi\mathrm{i}}\left[\frac{1}{2}\,\bar\partial_0\left(\mathfrak{m}^{\dot\beta}\,\frac{\partial\mathfrak{m}_{\dot\beta}}{\partial y^{\dot\alpha}}\right) - \frac{\bar\partial_0\mathfrak{m}_{\dot\alpha}}{\lambda_2}\right]\\ &= \int_Y \mathrm{D}\lambda\wedge\bar\delta(\lambda_2)\,\mathfrak{m}_{\dot\alpha}(y,\lambda) = -\mathfrak{m}_{\dot\alpha}(y,\kappa_1)\,.\end{aligned} \tag{3.65}$$

In getting from the second to the third line, we have dropped the total derivative term, integrated by parts in the second term, and recalled the definition $\bar e^0\,\bar\partial_0(1/\lambda_2) = 2\pi\mathrm{i}\,\bar\delta(\lambda_2)$ of a holomorphic delta function. The final integral can be done by going to the affine chart $\lambda_1 = 1$, $\lambda_2 = z$ on $\mathbb{P}^1$, in which $\mathrm{D}\lambda = \lambda_2\,\mathrm{d}\lambda_1 - \lambda_1\,\mathrm{d}\lambda_2 = -\mathrm{d}z$. This establishes the first of (3.63). The $\tilde y^{\dot\alpha}$-derivative can be computed in an analogous manner. □

[5] This calculation is explicitly described in the next chapter in its most general incarnation, in Eqs. (4.68)–(4.70).

Lemma 3.2 *Second derivatives of the twistor sigma model's on-shell action yield*

$$\left.\frac{\partial^2 S_h[m](y)}{\partial y^{\dot\alpha}\partial \tilde y^{\dot\beta}}\right|_{\text{on-shell}} = \epsilon_{\dot\alpha\dot\beta} + H_{\dot\alpha}{}^{\dot\gamma}(y,\kappa_1)\, H_{\dot\beta\dot\gamma}(y,\kappa_2)\,. \tag{3.66}$$

where $H^{\dot\alpha}{}_{\dot\beta}(y,\lambda) \in \mathrm{SL}(2,\mathbb{C})$ is the spin frame defined in (3.39).

Proof Using the integral equation (3.46), we can make the $y^{\dot\alpha}$-derivative more explicit:

$$\begin{aligned}\frac{\partial}{\partial y^{\dot\alpha}} S_h[m](y)\big|_{\text{on-shell}} &= -\frac{1}{2\pi \mathrm{i}}\int_Y \frac{\mathrm{D}\lambda}{\lambda_1\,\lambda_2^2}\wedge \mathcal{L}_{\partial_{\dot\alpha}} h|_Y \\ &= -\frac{1}{2\pi \mathrm{i}}\int_Y \frac{\mathrm{D}\lambda\wedge \bar e^0}{\lambda_1\,\lambda_2^2}\left[(\partial_{\dot\alpha}h_0 + h_{\dot\alpha})|_Y + \partial_{\dot\alpha}h_{\dot\beta}|_Y\left(F^{\dot\beta} - \frac{\bar\partial_0 \hat F^{\dot\beta}}{\langle\hat\lambda\,\lambda\rangle}\right)\right].\end{aligned} \tag{3.67}$$

Turning the crank, its $\tilde y^{\dot\beta}$-derivative can be massaged to yield a beautifully simple answer,

$$\left.\frac{\partial^2 S_h[m](y)}{\partial y^{\dot\alpha}\partial \tilde y^{\dot\beta}}\right|_{\text{on-shell}} = \frac{1}{2\pi \mathrm{i}}\int_Y \frac{\mathrm{D}\lambda\wedge \bar e^0}{\lambda_1\,\lambda_2^2}\left[\frac{\partial}{\partial \tilde y^{\dot\beta}} \lrcorner\, \bar\partial_0 \lrcorner\, p^*\mathrm{d}\theta_{\dot\alpha} - \bar\partial_0\left(\frac{\partial}{\partial \tilde y^{\dot\beta}}\lrcorner\, p^*\mathcal{L}_{\partial_{\dot\alpha}} h\right)\right]. \tag{3.68}$$

This has been simplified by means of $\mathrm{d}\theta^{\dot\alpha} = \theta^{\dot\gamma}\wedge \mathcal{L}_{\partial_{\dot\gamma}}\mathcal{L}_{\partial^{\dot\alpha}} h + \mathcal{L}_{\partial^{\dot\alpha}} T \bmod \mathrm{D}\lambda$, which follows from putting together Eqs. (3.20), (3.26) and (3.28) of the previous section. Once again, the second term in (3.68) is killed by Stokes' theorem on remembering the boundary conditions on h and the regularity of $F^{\dot\alpha}$ at κ_1, κ_2.

Now assume $T = 0$. At the end of the previous section, we used this to discover an $\mathrm{SL}(2,\mathbb{C})$-valued dotted spin frame $H^{\dot\alpha}{}_{\dot\beta}(y,\lambda)$. Its defining property (3.39) leads to

$$\left.\frac{\partial^2 S_h[m](y)}{\partial y^{\dot\alpha}\partial \tilde y^{\dot\beta}}\right|_{\text{on-shell}} = \frac{1}{2\pi \mathrm{i}}\frac{\partial}{\partial \tilde y^{\dot\beta}} \lrcorner\, \boldsymbol{\theta}^{\gamma\dot\gamma}\int_Y \frac{\mathrm{D}\lambda\wedge \bar e^0}{\lambda_1\,\lambda_2^2}\,\lambda_\gamma\,\bar\partial_0 H_{\dot\alpha\dot\gamma}\,, \tag{3.69}$$

where $\boldsymbol{\theta}^{\delta\dot\delta}$ is the space-time tetrad.

Using the equation of motion $\bar\partial_0 \lrcorner\, p^*\theta^{\dot\alpha} = 0$, we can show that

$$p^*\theta^{\dot\alpha} = \mathrm{d}_y F^{\dot\alpha} - p^*\partial^{\dot\alpha} h_{\dot\beta}\,\frac{\mathrm{d}_y \hat F^{\dot\beta}}{\langle\hat\lambda\,\lambda\rangle} \qquad \bmod \mathrm{D}\lambda\,. \tag{3.70}$$

From this, we can read off boundary values of the $(1,0)$-forms:

$$p^*\theta^{\dot\alpha}(y,\kappa_1) = \mathrm{d}y^{\dot\alpha}\,, \qquad p^*\theta^{\dot\alpha}(y,\kappa_2) = \mathrm{d}\tilde y^{\dot\alpha}\,, \tag{3.71}$$

having applied (3.48), coupled with the fact that $\partial^{\dot\alpha} h_{\dot\beta} = 0$ at κ_1, κ_2. Equating these to the right hand side of (3.39) at κ_1, κ_2 allows us to extract components of the tetrad,

$$\begin{aligned}\theta^{1\dot\alpha} &\equiv \kappa_{1\alpha}\,\theta^{\alpha\dot\alpha} = -H_{\dot\beta}{}^{\dot\alpha}(y,\kappa_1)\,\mathrm{d}y^{\dot\beta}\,,\\ \theta^{2\dot\alpha} &\equiv \kappa_{2\alpha}\,\theta^{\alpha\dot\alpha} = -H_{\dot\beta}{}^{\dot\alpha}(y,\kappa_2)\,\mathrm{d}\tilde y^{\dot\beta}\,,\end{aligned} \tag{3.72}$$

where the spin frame was inverted using $H^{\dot\alpha}{}_{\dot\beta}\,H_{\dot\alpha}{}^{\dot\gamma} = -\delta^{\dot\gamma}_{\dot\beta}$. In particular, $\partial/\partial\tilde y^{\dot\beta}$ only has a non-zero contraction with $\theta^{2\dot\gamma}$. Continuing with the computation of (3.69) finally yields

$$\begin{aligned}\left.\frac{\partial^2 S_h[m](y)}{\partial y^{\dot\alpha}\partial\tilde y^{\dot\beta}}\right|_{\text{on-shell}} &= -\frac{H_{\dot\beta}{}^{\dot\gamma}(y,\kappa_2)}{2\pi\mathrm{i}}\int_Y \frac{\mathrm{D}\lambda\wedge\bar e^0}{\lambda_1\lambda_2}\,\bar\partial_0 H_{\dot\alpha\dot\gamma}\\ &= H_{\dot\beta}{}^{\dot\gamma}(y,\kappa_2)\left(H_{\dot\alpha\dot\gamma}(y,\kappa_2) - H_{\dot\alpha\dot\gamma}(y,\kappa_1)\right)\\ &= \epsilon_{\dot\alpha\dot\beta} + H_{\dot\alpha}{}^{\dot\gamma}(y,\kappa_1)\,H_{\dot\beta\dot\gamma}(y,\kappa_2)\,,\end{aligned} \tag{3.73}$$

having integrated by parts to localize the λ_α integral at the two simple poles. □

Proof of Theorem 3.2: Second derivatives of the scalar Ω given in (3.62) now become

$$\frac{\partial^2\Omega}{\partial y^{\dot\alpha}\partial\tilde y^{\dot\beta}} = H_{\dot\alpha\dot\gamma}(y,\kappa_1)\,H_{\dot\beta}{}^{\dot\gamma}(y,\kappa_2)\,. \tag{3.74}$$

Hence, Ω satisfies the heavenly Eq. (3.13) because the spin frame is unimodular:

$$\begin{aligned}\frac{1}{2}\,\frac{\partial^2\Omega}{\partial y^{\dot\alpha}\partial\tilde y^{\dot\beta}}\,\frac{\partial^2\Omega}{\partial y_{\dot\alpha}\partial\tilde y_{\dot\beta}} &= \frac{1}{2}\,H_{\dot\alpha\dot\gamma}(y,\kappa_1)\,H^{\dot\alpha}{}_{\dot\delta}(y,\kappa_1)\,H_{\dot\beta}{}^{\dot\gamma}(y,\kappa_2)\,H^{\dot\beta\dot\delta}(y,\kappa_2)\\ &= \frac{1}{2}\,\epsilon_{\dot\gamma\dot\delta}\,\epsilon^{\dot\gamma\dot\delta} = 1\,.\end{aligned} \tag{3.75}$$

Lastly, we can compute the space-time metric

$$\begin{aligned}g = \theta^{\alpha\dot\alpha}\,\theta_{\alpha\dot\alpha} &= 2\,\theta^{2\dot\alpha}\,\theta^{1}{}_{\dot\alpha} = 2\,H_{\dot\alpha\dot\gamma}(y,\kappa_1)\,H_{\dot\beta}{}^{\dot\gamma}(y,\kappa_2)\,\mathrm{d}y^{\dot\alpha}\,\mathrm{d}\tilde y^{\dot\beta}\\ &= 2\,\frac{\partial^2\Omega}{\partial y^{\dot\alpha}\partial\tilde y^{\dot\beta}}\,\mathrm{d}y^{\dot\alpha}\,\mathrm{d}\tilde y^{\dot\beta}\,.\end{aligned} \tag{3.76}$$

Hence, Ω gives the Kähler scalar of space-time. □

Our expression (3.62) can be thought of as a Penrose integral formula for solutions of the fully non-linear heavenly equation. By treating $y^{\dot\alpha}$, $\tilde y^{\dot\alpha}$ as coordinates on complexified space-time $M_{\mathbb{C}}$, it gives rise to complexified hyperkähler metrics. A choice of involution j as in Theorem 3.1 allows one to pick a real slice, i.e., a relation between $\tilde y^{\dot\alpha}$ and the complex conjugates of $y^{\dot\alpha}$, yielding a Riemannian hyperkähler metric. Eg., on $\mathbb{R}^4$ one would set $\tilde y^{\dot\alpha} = \hat y^{\dot\alpha} = (-\overline{y^{\dot 2}}, \overline{y^{\dot 1}})$. Alternatively, we can continue to $(2,2)$ signature and find "real" hyperkähler metrics.

This shows the broader scope of twistor methods beyond linear field equations. Similar formulae can also be built for other descriptions of hyperkähler metrics like Plebanski's second heavenly equation, as well as for quaternionic metrics on 4-manifolds. This can also be generalized to $4k$ dimensional quaternionic geometry for $k \in \mathbb{Z}_{\geq 1}$. We discuss such generalizations in the next chapter, but this may be safely skipped on first reading. A more complete discussion can also be found in [3].

For the purposes of four-dimensional gravitational physics, we will continue to focus on the model (3.49). Its main practical application will be to the computation of graviton scattering amplitudes. We come to this in Chap. 4.

References

1. L. J. Mason and N. M. J. Woodhouse, *Integrability, selfduality, and twistor theory*. Oxford University Press, 1996
2. Plebanski JF (1975) Some solutions of complex Einstein equations. J Math Phys 16:2395–2402
3. Adamo T, Mason L, Sharma A (2021) Twistor sigma models for quaternionic geometry and graviton scattering. arXiv:2103.16984
4. Chakravarty S, Mason L, Newman E (1991) Canonical structures on antiselfdual four manifolds and the diffeomorphism group. J Math Phys 32:1458–1464
5. Dunajski M, Mason L (2000) HyperKahler hierarchies and their twistor theory. Commun Math Phys 213:641–672 (2000). www.math/0001008
6. Woodhouse NMJ (1985) Real methods in twistor theory. Class Quant Grav 2:257–291
7. Penrose R (1976) The nonlinear graviton. Gen Rel Grav 7:171–176
8. Atiyah MF, Hitchin NJ, Singer IM (1978) Selfduality in four-dimensional Riemannian geometry. Proc Roy Soc Lond A 362:425–461
9. Penrose R (1976) Nonlinear gravitons and curved twistor theory. Gen Rel Grav 7:31–52
10. Hitchin NJ, Karlhede A, Lindstrom U, Rocek M (1987) Hyperkahler metrics and supersymmetry. Commun. Math. Phys. 108:535
11. McDuff D, Salamon D (2012) J-Holomorphic Curves and Symplectic Topology. American Mathematical Society, 2nd edn
12. Capovilla R, Jacobson T, Dell J, Mason LJ (1991) Selfdual two forms and gravity. Class Quant Grav 8:41–57
13. Mason LJ, Skinner D (2010) Gravity, twistors and the MHV formalism. Commun. Math. Phys. 294:827–862 (2010). arXiv:0808.3907
14. Sharma A (2021) Twistor action for general relativity. arXiv:2104.07031

Chapter 4
Generalizations

In the previous chapter, we introduced a chiral sigma model into twistor space whose on-shell action computed Kähler scalars for the underlying self-dual vacuum 4-manifold. In this chapter, we will further elaborate on its geometric origins. This will clarify how it can compute the full Kähler scalar including the Kähler scalar of flat space, so that it may be understood covariantly instead of perturbatively around flat space. Following this, we will generalize our sigma model to higher dimensions, to other descriptions of hyperkähler spaces, as well as to quaternion-Kähler spaces. This last case corresponds to turning on a cosmological constant Λ.

The topics of this chapter will not be immediately needed for the chapters on graviton scattering that follow. So the reader interested in the physical applications of our models may feel free to skip it. Nevertheless, some ideas of this chapter have indeed made their way to physics in other surprising setups like strong coupling amplitudes of $\mathcal{N} = 4$ super Yang-Mills theory [1]. So it may be viewed as a starting point for future explorations.

4.1 Sigma Models from Symplectic Potentials

In this section, let us continue to work in four space-time dimensions. The generalization to $4k$ dimensions with $k \geq 1$ will be carried out in the next two sections.

We motivated the twistor sigma model (3.49) as a 2d CFT (with defects at $\lambda = \kappa_1, \kappa_2$) whose equations of motion govern the twistor curves $\mu^{\dot\alpha} = F^{\dot\alpha}(y, \lambda)$. But otherwise, it feels a bit miraculous that it is so directly related to Kähler scalars on space-time. In this section, we provide an understanding of how twistor sigma models can be constructed from the symplectic structure on twistor space. Since it is this latter structure that encodes the Ricci-flatness of the space-time metric, it demystifies the relation of twistor sigma models with the space-time geometry. It also goes part way toward expressing the sigma models in a covariant guise.

A. Sharma, *Twistor Sigma Models*, Springer Theses,
https://doi.org/10.1007/978-3-031-50751-9_4

In Eq. (3.19), we parametrized a complex structure deformation of $\mathbb{PT}$ by a set of deformed $(1,0)$-forms, and called the resulting almost complex manifold a curved twistor space $\mathbb{P}\mathscr{T}$. In what follows, it will prove slightly easier to work in complex coordinates on $\mathbb{PT}$. Let us work with the homogeneous coordinates $Z^A = (\mu^{\dot\alpha}, \lambda_\alpha)$ on $\mathbb{CP}^3$, as restricted to $\mathbb{PT} = \mathbb{CP}^3 - \{\lambda_\alpha = 0\}$, which are related to the (x, λ) coordinates via $\mu^{\dot\alpha} = x^{\alpha\dot\alpha}\lambda_\alpha$. In these coordinates, staying in the Ricci-flat case, we can describe our complex structure deformation by the deformed $(0,1)$-forms

$$\theta^{\dot\alpha} = \mathrm{d}\mu^{\dot\alpha} + \epsilon^{\dot\alpha\dot\beta}\mathcal{L}_{\partial_{\dot\beta}}h \tag{4.1}$$

keeping $e^0 = \mathrm{D}\lambda$ undeformed as usual, and viewing $\mathrm{d}\mu^{\dot\alpha}$ as the exterior derivative of a section of $\mathscr{O}(1)$ through the Chern connection $\mathrm{d}\mu^{\dot\alpha} = \mathrm{d}_{\mathbb{C}^4}\mu^{\dot\alpha} - \frac{\langle\hat\lambda\,\mathrm{d}\lambda\rangle}{\langle\hat\lambda\,\lambda\rangle}\,\mu^{\dot\alpha}$. These differ from the $\theta^{\dot\alpha}$ defined in (3.19) by factors proportional to $\mathrm{D}\lambda$, thereby defining an equivalent almost complex structure (when $V^0 = 0$). Additionally, the 2-form

$$\Sigma = \theta^{\dot\alpha} \wedge \theta_{\dot\alpha} \tag{4.2}$$

agrees with the 2-form (3.22) of the previous chapter mod $\mathrm{D}\lambda$. So when $T \equiv \bar\partial h + \frac{1}{2}\{h, h\} = 0$, it defines the same fiberwise symplectic structure.

Our first task is to find a symplectic potential for Σ. That is, we look for a potential ϑ such that $\mathrm{d}\vartheta = \Sigma$ mod $\mathrm{D}\lambda$ when $T = 0$. A natural guess that works out is

$$\vartheta = [\mu\,\mathrm{d}\mu] + 2h\,. \tag{4.3}$$

Indeed, computing the exterior derivative and "completing the square" yields [2]

$$\begin{aligned}
\mathrm{d}\vartheta &= \mathrm{d}\mu^{\dot\alpha} \wedge \mathrm{d}\mu_{\dot\alpha} + 2\,\mathrm{d}\mu^{\dot\alpha} \wedge \mathcal{L}_{\partial_{\dot\alpha}}h + 2\,\bar\partial h \qquad \text{mod } \mathrm{D}\lambda \\
&= (\theta^{\dot\alpha} + \mathcal{L}_{\partial^{\dot\alpha}}h) \wedge (\theta_{\dot\alpha} - \mathcal{L}_{\partial_{\dot\alpha}}h) + 2\,(\theta^{\dot\alpha} + \mathcal{L}_{\partial^{\dot\alpha}}h) \wedge \mathcal{L}_{\partial_{\dot\alpha}}h + 2\,\bar\partial h \qquad \text{mod } \mathrm{D}\lambda \\
&= \Sigma + 2T \qquad \text{mod } \mathrm{D}\lambda\,.
\end{aligned} \tag{4.4}$$

So $\mathrm{d}\vartheta = \Sigma$ precisely if T vanishes. This also gives another proof of the fact that $\mathrm{d}\Sigma = 0$ when $T = 0$. Note that ϑ in general only provides a local potential for Σ, and in particular, it is *not* a $(1,0)$-form. Nonetheless, it turns out to be the geometric object that most directly enters our sigma models.

To make the connection, we introduce a degree -1 version of the map $\mathbb{P}^1 \to \mathbb{PT}$ sending $\lambda_\alpha \mapsto \mu^{\dot\alpha} = F^{\dot\alpha}(y, \lambda)$. Introduce new homogeneous coordinates $\sigma_\alpha \in \mathbb{P}^1$ and relate them to $\lambda_\alpha \in \mathbb{P}^1$ by the degree -1 map

$$\lambda_\alpha(\sigma) = \frac{\sigma_\alpha}{\sigma_1\sigma_2} = \left(\frac{1}{\sigma_2}, \frac{1}{\sigma_1}\right). \tag{4.5}$$

In affine coordinates $z = \lambda_2/\lambda_1$, $\sigma = \sigma_2/\sigma_1$, this is just an inversion $\sigma = 1/z$. The inverse transformation is $\sigma_\alpha = \lambda_\alpha/\lambda_1\lambda_2$. The incidence relations may now be expressed as a map

$$\mu^{\dot\alpha}(\sigma) = F^{\dot\alpha}(y, \lambda(\sigma)) \tag{4.6}$$

of homogeneity degree -1 in σ_α. The usage of such degree -1 maps was spearheaded by Geyer et al. in work on ambitwistor strings in four dimensions [3].

The symplectic potential ϑ provides a natural means of writing a chiral sigma model governing degree -1 maps $\mathbb{P}^1 \to \mathbb{PT}$:

$$S = \frac{1}{4\pi \mathrm{i}} \int_{\mathbb{P}^1} \mathrm{D}\sigma \wedge \vartheta\,. \tag{4.7}$$

In this expression, ϑ is pulled back by the map $\mathbb{P}^1 \to \mathbb{PT}$. Its pullback has weight -2 in σ_α, so it has been integrated over $\mathbb{P}^1$ against the standard measure $\mathrm{D}\sigma = \sigma^\alpha \mathrm{d}\sigma_\alpha$. Note that this action is covariant, because ϑ may be thought of as an abstract symplectic potential for a fiberwise symplectic structure Σ on an arbitrary curved twistor space $\mathbb{P}\mathscr{T}$. That is, $\mathbb{P}\mathscr{T}$ a priori need not be expressible as a deformation of flat twistor space $\mathbb{PT}$. The action S is still understood to be a functional of the map $\mathbb{P}^1 \to \mathbb{PT}$, which can be made explicit by any good choice of coordinates on $\mathbb{P}\mathscr{T}$.

Of course, we are working with the explicit potential (4.3), in which case this action expands out to

$$S[\mu] = \frac{1}{2\pi \mathrm{i}} \int_{\mathbb{P}^1} \mathrm{D}\sigma \wedge \left(\frac{1}{2}\,[\mu\,\bar\partial\mu] + h(\mu, \lambda)\right) \tag{4.8}$$

where $\bar\partial = \mathrm{d}\hat\sigma \cdot \partial_{\hat\sigma}$. Since h has homogeneity $+2$ in $Z = (\mu, \lambda)$, it has homogeneity -2 in σ_α as desired. This action already resembles the twistor sigma model (3.49) to a great degree (no pun intended), except that we are using homogeneity degree -1 fields as its fundamental variables. But it is not quite (3.49), as it does not a priori know about the choice of coordinates $y^{\alpha\dot\alpha} = (y^{\dot\alpha}, \tilde y^{\dot\alpha})$ on space-time M. Only the specific map $\mu^{\dot\alpha}(\sigma) = F^{\dot\alpha}(y, \lambda(\sigma))$ given by the incidence relations knew about these coordinates.

To relate the two, one adds a source term to (4.8) to define the new action

$$S_h[\mu](y, \tilde y) = \frac{1}{2\pi \mathrm{i}} \int_{\mathbb{P}^1} \mathrm{D}\sigma \wedge \left(\frac{1}{2}\,[\mu\,\bar\partial\mu] + h(\mu, \lambda) - \mu^{\dot\alpha} J_{\dot\alpha}\right) \tag{4.9}$$

with $J_{\dot\alpha}$ chosen to be a sum of two point-like sources

$$J^{\dot\alpha} = 2\pi \mathrm{i}\, y^{\dot\alpha}\, \bar\delta(\sigma_2) + 2\pi \mathrm{i}\, \tilde y^{\dot\alpha}\, \bar\delta(\sigma_1)\,. \tag{4.10}$$

This is the natural guess, because $y^{\dot\alpha}$, $\tilde y^{\dot\alpha}$ originally acted as boundary conditions for the degree 1 incidence relations in (3.48). The sources now insert the required defects into the theory. A covariantization of this source term has not been developed as of the writing of this work.

To understand the effect of the sources, complete the square to find

$$S_h = \frac{1}{2\pi\mathrm{i}}\int_{\mathbb{P}^1} \mathrm{D}\sigma \wedge \left\{\frac{1}{2}\,(\mu^{\dot\alpha} - \bar\partial^{-1}J^{\dot\alpha})\,\bar\partial(\mu_{\dot\alpha} - \bar\partial^{-1}J_{\dot\alpha}) - \frac{1}{2}\,J_{\dot\alpha}\bar\partial^{-1}J^{\dot\alpha} + h(\mu,\lambda)\right\}. \tag{4.11}$$

Introducing a new field $\mathrm{m}^{\dot\alpha}(\sigma)$ of homogeneity -1 in σ_α by the field redefinition

$$\mu^{\dot\alpha}(\sigma) = \bar\partial^{-1}J^{\dot\alpha} + \mathrm{m}^{\dot\alpha}(\sigma) = \frac{y^{\dot\alpha}}{\sigma_2} + \frac{\tilde y^{\dot\alpha}}{\sigma_1} + \mathrm{m}^{\dot\alpha}(\sigma), \tag{4.12}$$

one can reduce (4.11) to

$$S_h = \frac{1}{4\pi\mathrm{i}}\int_{\mathbb{P}^1} \mathrm{D}\sigma \wedge J^{\dot\alpha}\bar\partial^{-1}J_{\dot\alpha} + \frac{1}{2\pi\mathrm{i}}\int_{\mathbb{P}^1} \mathrm{D}\sigma \wedge \left\{\frac{1}{2}\,\mathrm{m}^{\dot\alpha}\bar\partial\mathrm{m}_{\dot\alpha} + h\left(\frac{y^{\dot\alpha}}{\sigma_2} + \frac{\tilde y^{\dot\alpha}}{\sigma_1} + \mathrm{m},\lambda\right)\right\}. \tag{4.13}$$

The σ integral in the first term can be localized against the delta functions in $J^{\dot\alpha}$. Up to an overall sign, it generates precisely the flat space Kähler scalar:

$$\frac{1}{4\pi\mathrm{i}}\int_{\mathbb{P}^1} \mathrm{D}\sigma \wedge J^{\dot\alpha}\bar\partial^{-1}J_{\dot\alpha} = -\epsilon_{\dot\alpha\dot\beta}\,y^{\dot\alpha}\tilde y^{\dot\beta} = [y\,\tilde y]\,. \tag{4.14}$$

The second term can finally be converted into the familiar twistor sigma model by setting $\sigma_\alpha = \lambda_\alpha/\lambda_1\lambda_2$ and rewriting it in terms of the homogeneity $+1$ field

$$m^{\dot\alpha}(\lambda) := \lambda_1\lambda_2\mathrm{m}^{\dot\alpha}(\lambda) = \mathrm{m}^{\dot\alpha}\left(\frac{\lambda_\alpha}{\lambda_1\lambda_2}\right) = \mathrm{m}^{\dot\alpha}(\sigma)\,. \tag{4.15}$$

The measure changes as $\mathrm{D}\sigma = \mathrm{D}\lambda/\lambda_1^2\lambda_2^2$, the dbar operator changes covariantly to $\bar\partial = \mathrm{d}\hat\lambda\cdot\partial_{\hat\lambda}$ (the $\hat\lambda_1\hat\lambda_2 = 0$ poles in $\mathrm{d}\hat\sigma_\alpha$ cancel against the zeroes of $\partial_{\hat\sigma_\alpha}$), and the action reduces to

$$S_h = [y\,\tilde y] + \frac{1}{2\pi\mathrm{i}}\int_{\mathbb{P}^1} \frac{\mathrm{D}\lambda}{\lambda_1^2\lambda_2^2} \wedge \left\{\frac{1}{2}\,m^{\dot\alpha}\bar\partial m_{\dot\alpha} + h(\lambda_1 y + \lambda_2\tilde y + m,\lambda)\right\}. \tag{4.16}$$

Since $F^{\dot\alpha}(y,\lambda) = \lambda_1 y^{\dot\alpha} + \lambda_2\tilde y^{\dot\alpha} + m^{\dot\alpha}$ was precisely our deformed incidence relation, we see that our sources have generated this deformation dynamically.

Taking (4.16) on-shell and recalling Theorem 3.2, one discovers that $S_h[\mu]$ computes the total Kähler potential of the metric on M,

$$\Omega(y,\tilde y) = -S_h[\mu](y,\tilde y)\Big|_{\text{on-shell}}\,. \tag{4.17}$$

It automatically contains the flat space part $\epsilon_{\dot{\alpha}\dot{\beta}} y^{\dot{\alpha}} \tilde{y}^{\dot{\beta}}$. This is consistent with the covariant nature of our starting point (4.7). This analysis also clarifies the role of the defects: they act as source terms that fix a choice of complex coordinates $y^{\dot{\alpha}}$, $\tilde{y}^{\dot{\alpha}}$ on M in terms of which the metric is Kähler.

In the coming sections, we will see that the symplectic-potential-based action (4.7) forms the "universal part" of our sigma models, and the choice of sources/defects changes from application to application. We remark that interpreting defects as source terms is again motivated from the study of worldsheet correlators of the 4d ambitwistor string [3]. In that setting, the sources are not put in by hand, but enter through vertex operators representing single-particle states on space-time. We suspect that looking for the origin of our sources—perhaps along the lines of ambitwistor strings—will help us discover a generalization of our sigma models that goes beyond the self-dual sector.

A parallel open question is to look for supersymmetric generalizations, as supersymmetry might allow us to construct covariant space-time quantities that are independent of the defect locations. It might allow us to directly construct the hyperkähler metric or the triad of ASD 2-forms $\Sigma^{\alpha\beta}$ on space-time, instead of Kähler scalars adapted to specific complex coordinates. Indeed, this is reminiscent of the construction of hyperkähler metrics through nonlinear sigma models with $\mathcal{N} = 4$ supersymmetry [4], with twistor space being highly analogous to harmonic superspace [5].

4.2 Higher Dimensions and Heavenly Equations

We now describe the generalizations of our sigma models to higher dimensions. We revisit the first heavenly equation in this context, as well as construct sigma models that compute solutions of the second heavenly equation.

Higher dimensional hyperkähler geometry. Let (M, g) be a $4k$ dimensional hyperkähler manifold, with $k \in \mathbb{Z}_{\geq 1}$. The hyperkähler condition states that the holonomy of the Levi-Civita connection of g lies in $\mathrm{Sp}(k)$. Decomposing the cotangent spaces as representations of $\mathrm{Sp}(1) \times \mathrm{Sp}(k) \subset \mathrm{SO}(4k)$, we can introduce "spinor indices" $\alpha = 1, 2$ and $\dot{\alpha} = \dot{1}, \dot{2}, \ldots, \dot{2k}$ so as to construct spinor equivalents $\boldsymbol{\theta}^{\alpha\dot{\alpha}}$ of a given coframe (vielbein) $\boldsymbol{\theta}^a$ for g. Let $\epsilon_{\alpha\beta}$ and $\epsilon_{\dot{\alpha}\dot{\beta}}$ be the invariant symplectic matrices of $\mathrm{Sp}(1)$ and $\mathrm{Sp}(k)$ respectively, which can be used to raise and lower the spinor indices. In terms of these, the metric is expressed in the usual format

$$g = \epsilon_{\alpha\beta}\epsilon_{\dot{\alpha}\dot{\beta}}\boldsymbol{\theta}^{\alpha\dot{\alpha}}\boldsymbol{\theta}^{\beta\dot{\beta}} . \tag{4.18}$$

Because $\mathrm{Sp}(1) = \mathrm{SU}(2)$, this reduces to the standard spinor notation when $k = 1$. Spinor components of various tensors may now be defined as components in this coframe and its dual. As in four dimensions, holonomy reduction is the statement that the Riemann curvature 2-form takes the form

$$R_{\alpha\dot{\alpha}\beta\dot{\beta}} = \epsilon_{\alpha\beta}\,\tilde{R}_{\dot{\alpha}\dot{\beta}}\,. \tag{4.19}$$

That is, it is as an element of the Lie algebra of Sp(k).

A hyperkähler manifold can equivalently be defined in terms of its complex structures. (M, g) comes equipped with a triplet of covariantly constant complex structures I, J, K satisfying the quaternion algebra

$$I^2 = J^2 = K^2 = IJK = -1\,. \tag{4.20}$$

A useful review of this may be found in [6]. In fact, there is an entire 2-sphere's worth of covariantly constant complex structures, because each linear combination $aI + bJ + cK$ such that $a^2 + b^2 + c^2 = 1$ is again covariantly constant and satisfies $(aI + bJ + cK)^2 = -1$. The metric g is Kähler with respect to each of these, though the Kähler potentials vary as we vary the complex structure.

For such a hyperkähler manifold, we can construct a twistor space $\mathbb{P}\mathscr{T}$ as the bundle of pointwise almost complex structures that are compatible with the metric and a choice of orientation. Alternatively, it can be viewed as a bundle of 2-spheres $S^2 \to \mathbb{P}\mathscr{T} \to M$ whose global sections are precisely the complex structures $aI + bJ + cK$ described above. Just as in the four-dimensional case, $\mathbb{P}\mathscr{T}$ admits the structure of a complex manifold of complex dimension $2k + 1$; the details of this construction have been worked out in [4].

One can again show that $\mathbb{P}\mathscr{T}$ has the structure of a holomorphic fiber bundle over $\mathbb{P}^1 \simeq S^2$, and sections of this fibration can be identified with the $\mathbb{P}^1$ fibers of the bundle of almost complex structures. We say that each such $\mathbb{P}^1$ is a twistor curve, and they can all be checked to have normal bundle $\mathbb{C}^{2k} \otimes \mathscr{O}(1)$. The antipodal map on each $\mathbb{P}^1$ induces an antiholomorphic involution on $\mathbb{P}\mathscr{T}$. In reverse, given such an involution, one can select "real" twistor curves as those that are mapped to themselves by this involution. One can recover a complexification $M_{\mathbb{C}}$ of the underlying hyperkähler manifold M as the space of all sections of $\mathbb{P}\mathscr{T} \to \mathbb{P}^1$, whereas the space of sections constituting real twistor curves recovers the real slice M.

Each complex structure I, J, K can be used to define a Kähler form for the hyperkähler metric g, which we can denote by $\omega_I, \omega_J, \omega_K$. For example, $\omega_I(v, w) = g(Iv, w)$ for vectors v, w, and so on. These may be put together into a bispinor of closed 2-forms $\Sigma^{\alpha\beta}$ with components

$$\Sigma^{11} = \omega_I + \mathrm{i}\,\omega_J\,, \qquad \Sigma^{12} = \mathrm{i}\,\omega_K\,, \qquad \Sigma^{22} = \overline{\Sigma^{11}}\,. \tag{4.21}$$

It can be checked that Σ^{11} is holomorphic with respect to K (see Eq. (3.34) of [4]). In terms of the coframe $\theta^{\alpha\dot{\alpha}}$ employed in holonomy reduction, these can be summarized through the familiar formula

$$\Sigma^{\alpha\beta} = \boldsymbol{\theta}^{\alpha\dot{\alpha}} \wedge \boldsymbol{\theta}^{\beta}{}_{\dot{\alpha}}\,. \tag{4.22}$$

If λ_α denote homogeneous coordinates on the $\mathbb{P}^1$ fibers of $\mathbb{P}\mathscr{T} \to M$, we can again define a 2-form

$$\Sigma = \lambda_\alpha \lambda_\beta \Sigma^{\alpha\beta} \,. \tag{4.23}$$

This can be shown to define a holomorphic symplectic form on each fiber of $\mathbb{P}\mathscr{T} \to \mathbb{P}^1$, i.e., at each fixed λ_α slice of $\mathbb{P}\mathscr{T}$.

The twistor sigma model again provides a formulaic way of running this construction in reverse. We will start with a fiberwise symplectic structure Σ on $\mathbb{P}\mathscr{T}$ and construct Kähler potentials for the hyperkähler metric g for any choice of complex structure $aI + bJ + cK$. Note that each choice of complex structure can equivalently be parametrized by a choice of spinors $\lambda = \kappa_1, \kappa_2$ such that $\kappa_2 = \hat{\kappa}_1$, with κ_1 being homogeneous coordinates of the point $(a, b, c) \in S^2$, and κ_2 being its antipode and representing the conjugate complex structure $-(aI + bJ + cK)$. We will again keep κ_1, κ_2 arbitrary for simplicity of notation, and also because this simply corresponds to working in the complexified setting.

There are two types of potentials for hyperkähler metrics that we can construct using the twistor sigma models. The first is the familiar Kähler potential Ω in say the complex structure K, in terms of which the metric may be expressed as

$$g = 2\,\Omega_{\dot\alpha\dot\beta}\,\mathrm{d}y^{\dot\alpha}\,\mathrm{d}\tilde{y}^{\dot\beta}\,, \qquad \Omega_{\dot\alpha\dot\beta} \equiv \frac{\partial^2\Omega}{\partial y^{\dot\alpha}\partial\tilde{y}^{\dot\beta}}\,. \tag{4.24}$$

Here, $y^{\dot\alpha}$ and $\tilde{y}^{\dot\alpha}$ denote a choice of complex coordinates with respect to K. Again, we can take $\tilde{y}^{\dot\alpha} = \hat{y}^{\dot\alpha} \equiv (\epsilon \cdot \bar{y})^{\dot\alpha}$ as the conjugation, where ϵ acts as a $2k \times 2k$ Levi-Civita tensor, but it is easier and more general to work with complexified coordinates $y^{\dot\alpha}, \tilde{y}^{\dot\alpha}$ for simplicity. The easiest choice of coframe to work with (albeit not obeying any reality condition) is

$$\boldsymbol{\theta}^{1\dot\alpha} = \mathrm{d}y^{\dot\alpha}\,, \qquad \boldsymbol{\theta}^{2\dot\alpha} = \Omega^{\dot\alpha\dot\beta}\mathrm{d}\tilde{y}_{\dot\beta}\,. \tag{4.25}$$

The resulting 2-forms $\Sigma^{\alpha\beta}$ are found to be

$$\Sigma^{11} = \mathrm{d}y^{\dot\alpha} \wedge \mathrm{d}y_{\dot\alpha}\,, \qquad \Sigma^{12} = -\Omega_{\dot\alpha\dot\beta}\,\mathrm{d}y^{\dot\alpha} \wedge \mathrm{d}\tilde{y}^{\dot\beta}\,, \qquad \Sigma^{22} = \Omega^{\dot\alpha}{}_{\dot\beta}\Omega_{\dot\alpha\dot\gamma}\,\mathrm{d}\tilde{y}^{\dot\beta} \wedge \mathrm{d}\tilde{y}^{\dot\gamma}\,. \tag{4.26}$$

We can now study the condition $\mathrm{d}\Sigma^{\alpha\beta} = 0$ under which g is Ricci-flat.

The 2-forms Σ^{11}, Σ^{12} are seen to be closed by inspection. Following Plebanski [7], we impose $\mathrm{d}\Sigma^{22} = 0$ by demanding that $\tilde{y}^{\dot\beta}$ be Darboux coordinates for Σ^{22}, which is also consistent with the Euclidean reality condition $\Sigma^{22} = \overline{\Sigma^{11}}$. This yields a set of PDEs constituting a higher dimensional version of Plebanski's *first heavenly equation*,

$$\epsilon^{\dot\alpha\dot\beta}\Omega_{\dot\alpha\dot\gamma}\Omega_{\dot\beta\dot\delta} = \epsilon_{\dot\gamma\dot\delta}\,. \tag{4.27}$$

When these hold, one readily verifies from (4.26) that

$$\Sigma^{11} = \mathrm{d}y^{\dot\alpha} \wedge \mathrm{d}y_{\dot\alpha}\,, \qquad \Sigma^{12} = -\partial\tilde\partial\Omega\,, \qquad \Sigma^{22} = \mathrm{d}\tilde{y}^{\dot\alpha} \wedge \mathrm{d}\tilde{y}_{\dot\alpha}\,, \tag{4.28}$$

where $\partial = \mathrm{d}y \cdot \partial_y$ and $\tilde{\partial} = \mathrm{d}\tilde{y} \cdot \partial_{\tilde{y}}$ are the holomorphic and antiholomorphic exterior derivatives on M. Hence, we find a triplet of closed 2-forms, expressed in Darboux coordinates, using which one can reconstruct the rest of the hyperkähler structure.

The second type of potential is what we will call a *Plebanski scalar*, introduced by Plebanski in the four-dimensional context in [7]. The Kähler description came adapted to Darboux coordinates that put Σ^{11}, Σ^{22} in standard form. The Plebanski description follows by instead looking for Darboux coordinates $y^{\dot\alpha}$, $w^{\dot\alpha}$ in which Σ^{11} and Σ^{12} take the standard form

$$\Sigma^{11} = \mathrm{d}y^{\dot\alpha} \wedge \mathrm{d}y_{\dot\alpha}\,, \qquad \Sigma^{12} = \mathrm{d}y^{\dot\alpha} \wedge \mathrm{d}w_{\dot\alpha}\,. \tag{4.29}$$

This may be obtained from (4.28) by taking $y^{\dot\alpha}$ to be the same coordinate as before, and taking $w^{\dot\alpha}$ to be defined by $w^{\dot\alpha} = \partial\Omega/\partial y_{\dot\alpha}$.

Comparing with the definition (4.22) shows that the most general coframe that can give rise to these Σ^{11}, Σ^{12} is

$$\theta^{1\dot\alpha} = \mathrm{d}y^{\dot\alpha}\,, \qquad \theta^{2\dot\alpha} = \mathrm{d}w^{\dot\alpha} + \Theta^{\dot\alpha\dot\beta}\mathrm{d}y_{\dot\beta}\,, \qquad \Theta_{\dot\alpha\dot\beta} = \Theta_{\dot\beta\dot\alpha}\,. \tag{4.30}$$

This results in an a priori non-closed Σ^{22}:

$$\Sigma^{22} = \mathrm{d}w^{\dot\alpha} \wedge \mathrm{d}w_{\dot\alpha} + 2\,\Theta_{\dot\alpha\dot\beta}\,\mathrm{d}y^{\dot\alpha} \wedge \mathrm{d}w^{\dot\beta} + \Theta^{\dot\alpha}{}_{\dot\beta}\Theta_{\dot\alpha\dot\gamma}\,\mathrm{d}y^{\dot\beta} \wedge \mathrm{d}y^{\dot\gamma}\,. \tag{4.31}$$

The $\mathrm{d}y^{\dot\gamma} \wedge \mathrm{d}w^{\dot\alpha} \wedge \mathrm{d}w^{\dot\beta}$ component of the closure condition $\mathrm{d}\Sigma^{22} = 0$ leads to the constraint $\partial\Theta_{\dot\beta\dot\gamma}/\partial w^{\dot\alpha} = \partial\Theta_{\dot\alpha\dot\gamma}/\partial w^{\dot\beta}$, which can be locally solved by introducing the Plebanski scalar potential $\Theta(y, w)$ in terms of which

$$\Theta_{\dot\alpha\dot\beta} = \frac{\partial^2\Theta}{\partial w^{\dot\alpha}\partial w^{\dot\beta}}\,. \tag{4.32}$$

The $\mathrm{d}y^{\dot\alpha} \wedge \mathrm{d}y^{\dot\beta} \wedge \mathrm{d}w^{\dot\gamma}$ component of $\mathrm{d}\Sigma^{22} = 0$ then gives the equation of motion of Θ,

$$\frac{\partial^2\Theta}{\partial y^{[\dot\alpha}\partial w^{\dot\beta]}} + \frac{1}{2}\,\Theta^{\dot\gamma}{}_{[\dot\alpha}\Theta_{\dot\beta]\dot\gamma} = 0\,. \tag{4.33}$$

This is known as Plebanski's *second heavenly equation*. Solving this, one can use the vielbein (4.30) to construct the hyperkähler metric in a complexified form,

$$g = 2\,\mathrm{d}y_{\dot\alpha}\,\mathrm{d}w^{\dot\alpha} + 2\,\Theta_{\dot\alpha\dot\beta}\,\mathrm{d}y^{\dot\alpha}\,\mathrm{d}y^{\dot\beta}\,. \tag{4.34}$$

In general, the Kähler potential form leads to type D metrics in four dimensions, whereas the Plebanski potential form generates type N metrics.

Twistor sigma models. Next we describe our results about the twistor sigma models that pertain to higher dimensions and the Plebanski scalar. We will set up the twistor spaces as deformations of a flat twistor space, and build sigma models governing holomorphic rational curves. The curious reader is encouraged to peruse [8] for further details.

The twistor space of $\mathbb{R}^{4k}$ equipped with its flat metric is $\mathbb{PT} = \mathbb{P}^{2k+1} - \mathbb{P}^{2k-1}$. Here, one endows $\mathbb{P}^{2k+1}$ with homogeneous coordinates $Z^A = (\mu^{\dot{\alpha}}, \lambda_\alpha)$ and removes the locus $\lambda_\alpha = 0$ which is a copy of $\mathbb{P}^{2k-1}$. This converts $\mathbb{PT}$ into the total space of the holomorphic vector bundle $\mathbb{C}^{2k} \otimes \mathscr{O}(1) \to \mathbb{P}^1$. There is a 4k-complex-parameter family of sections of this bundle that recovers the complexification $\mathbb{C}^{4k}$ of the underlying hyperkähler manifold $\mathbb{R}^{4k}$, and sections invariant under the antiholomorphic involution

$$\lambda_\alpha \mapsto \hat{\lambda}_\alpha\,, \quad \mu^{\dot{\alpha}} \mapsto \hat{\mu}^{\dot{\alpha}} = (\epsilon \cdot \bar{\mu})^{\dot{\alpha}} \tag{4.35}$$

are in one-to-one correspondence with points of the real slice $\mathbb{R}^{4k}$.

$\mathbb{PT}$ comes equipped with the fiberwise symplectic structure $\mathrm{d}\mu^{\dot{\alpha}} \wedge \mathrm{d}\mu_{\dot{\alpha}}$. Deforming the complex structure of $\mathbb{PT}$ by hamiltonian deformations

$$\mathrm{D}\lambda \mapsto \mathrm{D}\lambda\,, \qquad \mathrm{d}\mu^{\dot{\alpha}} \mapsto \theta^{\dot{\alpha}} = \mathrm{d}\mu^{\dot{\alpha}} + \epsilon^{\dot{\alpha}\dot{\beta}}\mathcal{L}_{\partial_{\dot{\beta}}}h \tag{4.36}$$

results in twistor spaces of curved hyperkähler manifolds as long as the deformation is integrable, i.e., satisfies the vanishing of the torsion

$$T = \bar{\partial}h + \frac{1}{2}\{h, h\}\,, \qquad \{h, h\} := \epsilon^{\dot{\alpha}\dot{\beta}}\mathcal{L}_{\partial_{\dot{\alpha}}}h \wedge \mathcal{L}_{\partial_{\dot{\beta}}}h\,. \tag{4.37}$$

When this holds, one can construct a deformed fiberwise symplectic structure

$$\Sigma = \theta^{\dot{\alpha}} \wedge \theta_{\dot{\alpha}} \tag{4.38}$$

which descends to $\Sigma = \lambda_\alpha\lambda_\beta\Sigma^{\alpha\beta}$ on the new $4k$-dimensional manifold constructed as the space of deformed twistor curves. These are exactly the same expressions as in 4d. Only the index range of the dotted indices have changed.

We continue to denote by M the hyperkähler manifold obtained from the deformation h. Let $\mu^{\dot{\alpha}} = F^{\dot{\alpha}}(y, \lambda)$ be the deformed holomorphic curves obtained by solving the higher dimensional counterpart of the PDE (3.33). y here denotes some abstract set of coordinates on M. The two heavenly descriptions of hyperkähler geometry are found by defining complex coordinates on M in different ways. To get the Kähler scalar, one picks the coordinates as the boundary values of the incidence relations at two distinct points on the base $\mathbb{P}^1$:

$$y^{\dot{\alpha}} = F^{\dot{\alpha}}(y, \kappa_1)\,, \qquad \tilde{y}^{\dot{\alpha}} = F^{\dot{\alpha}}(y, \kappa_2)\,. \tag{4.39}$$

This is familiar from Sect. 3.3. On the other hand, the Plebanski scalar description is constructed by taking the coordinates to be boundary values of $F^{\dot{\alpha}}$ and its derivative at a single point:

$$y^{\dot\alpha} = F^{\dot\alpha}(y, \kappa_1)\,, \qquad w^{\dot\alpha} = \frac{\partial F^{\dot\alpha}}{\partial \lambda_2}(y, \kappa_1)\,. \tag{4.40}$$

Viewing these as boundary conditions, we can parametrize the incidence relations as

$$\mu^{\dot\alpha} = y^{\alpha\dot\alpha}\lambda_\alpha + \breve{\omega}\, \mathfrak{m}^{\dot\alpha}(y, \lambda)\,, \tag{4.41}$$

where $\mathfrak{m}^{\dot\alpha}$ is the degree -1 perturbation away from the flat incidence relation $\mu^{\dot\alpha} = y^{\alpha\dot\alpha}\lambda_\alpha$. The latter has been written in terms of a "double null" coordinate system that is given by $y^{\alpha\dot\alpha} = (y^{\dot\alpha}, \tilde{y}^{\dot\alpha})$ in the Kähler case and $y^{\alpha\dot\alpha} = (y^{\dot\alpha}, w^{\dot\alpha})$ in the Plebanski case. The quantity $\breve{\omega}$ that distinguishes the two cases reads

$$\breve{\omega} = \begin{cases} \lambda_1\lambda_2\,, & \text{Kähler} \\ \lambda_2^2\,, & \text{Plebanski} \end{cases} \tag{4.42}$$

where $\lambda_1 = \langle 2\,\lambda\rangle$, $\lambda_2 = \langle\lambda\,1\rangle$ and $\langle 1\,2\rangle = -1$ as before. In the literature, $\breve{\omega}$ is sometimes known as Pontecorvo's quadric [9] and is ultimately equivalent to a choice of complex structure on M to write the Kähler/Plebanski scalar with respect to.

For each of these two cases, we can introduce a twistor sigma model through the following set of steps. Start by defining a new $\mathbb{P}^1$-valued variable

$$\sigma_\alpha = \frac{\lambda_\alpha}{\breve{\omega}} \tag{4.43}$$

which will play the role of "worldsheet coordinate". In the Kähler case, this was given by $\sigma_\alpha = (1/\lambda_2, 1/\lambda_1)$, whereas in the the Plebanski case it becomes $\sigma_\alpha = (\lambda_1/\lambda_2^2, 1/\lambda_2)$. The action for the twistor sigma model is found by simply extending (4.9) to higher dimensions

$$S_h[\mu](y, \tilde{y}) = \frac{1}{2\pi \mathrm{i}} \int_{\mathbb{P}^1} \mathrm{D}\sigma \wedge \left(\frac{1}{2}\,[\mu\,\bar\partial\mu] + h(\mu, \lambda) - \mu^{\dot\alpha} J_{\dot\alpha}\right) \tag{4.44}$$

where $(\mu^{\dot\alpha}, \lambda_\alpha)$ are now the coordinates of $\mathbb{P}^{2k+1}$.

The choice of source $J^{\dot\alpha}$ is dictated by the flat space incidence relations:

$$\begin{aligned} J^{\dot\alpha} &= \mathrm{d}\bar\sigma \cdot \partial_{\bar\sigma}(y^{\alpha\dot\alpha}\lambda_\alpha) \\ &= \begin{cases} 2\pi \mathrm{i}\, y^{\dot\alpha}\,\bar\delta(\sigma_2) + 2\pi \mathrm{i}\,\tilde{y}^{\dot\alpha}\,\bar\delta(\sigma_1)\,, & \text{Kähler} \\ -2\pi \mathrm{i}\, w^{\dot\alpha}\sigma_1\bar\delta'(\sigma_2) + 2\pi \mathrm{i}\, w^{\dot\alpha}\,\bar\delta(\sigma_2)\,, & \text{Plebanski} \end{cases} \end{aligned} \tag{4.45}$$

having used $y^{\alpha\dot\alpha} = (y^{\dot\alpha}, \tilde{y}^{\dot\alpha})$ in the Kähler case and $y^{\alpha\dot\alpha} = (y^{\dot\alpha}, w^{\dot\alpha})$ in the Plebanski case. With this source, the field redefinition

$$\mu^{\dot\alpha}(\sigma) = \bar\partial^{-1} J^{\dot\alpha} + \mathfrak{m}^{\dot\alpha}(\sigma) \tag{4.46}$$

reproduces the structure of the incidence relations (4.41) upon setting $\sigma_\alpha = \lambda_\alpha/\check{\omega}$. We can combine this with the further relabelling

$$m^{\dot\alpha}(\lambda) := \mathrm{m}^{\dot\alpha}(\sigma) = \check{\omega}\, \mathrm{m}^{\dot\alpha}(\lambda) \tag{4.47}$$

to integrate out the source. The resulting action is

$$S_h[\mu] = \frac{1}{4\pi\mathrm{i}} \int_{\mathbb{P}^1} \mathrm{D}\sigma \wedge J^{\dot\alpha}\bar\partial^{-1} J_{\dot\alpha} + S_h[m]\,, \tag{4.48}$$

with the action for $m^{\dot\alpha}$ being given by

$$S_h[m] = \frac{1}{2\pi\mathrm{i}} \int_{\mathbb{P}^1} \frac{\mathrm{D}\lambda}{\check{\omega}^2} \wedge \left(\frac{1}{2}\,[m\,\bar\partial m] + h(F,\lambda)\right) \tag{4.49}$$

Here $F^{\dot\alpha} = y^{\alpha\dot\alpha}\lambda_\alpha + m^{\dot\alpha}$ as before, and we have used $\mathrm{D}\sigma = \mathrm{D}\lambda/\check{\omega}^2$.

In the Kähler case, the $J \cdot \bar\partial^{-1} J$ term reduces to minus the Kähler scalar of $\mathbb{R}^{4k}$, i.e, $-\epsilon_{\dot\alpha\dot\beta}y^{\dot\alpha}\tilde{y}^{\dot\beta}$. In the Plebanski case, it reduces to $w^{\dot\alpha}y_{\dot\alpha}$ times a coordinate independent divergence. But since it is linear in $w^{\dot\alpha}$, its second derivatives in $w^{\dot\alpha}$ are well-defined, which is all we need for our main theorem:

Theorem 4.1 *Fix the complex structure on M to be the one associated to $\kappa_1, \kappa_2 \in \mathbb{P}^1$ (where $\kappa_2 = \hat{\kappa}_1$ in the Euclidean signature case). Take h to satisfy $T = 0$ and suppose that $h = \check{\omega}^2\varphi$ for φ some smooth section of $\mathcal{O}(-2)$. For $\check{\omega} = \lambda_1\lambda_2$, the on-shell action of the sigma model* (4.48) *computes a Kähler scalar on M*

$$\Omega(y,\tilde{y}) = -S_h[\mu]\big|_{\text{on-shell}}\,, \qquad \check{\omega} = \lambda_1\lambda_2 \tag{4.50}$$

that solves Plebanski's first heavenly equation. For $\check{\omega} = \lambda_2^2$, it instead computes a solution of Plebanski's second heavenly equation

$$\Theta(y,w) = -S_h[\mu]\big|_{\text{on-shell}}\,, \qquad \check{\omega} = \lambda_2^2\,. \tag{4.51}$$

Since the metric (4.34) only depends on $\Theta_{\dot\alpha\dot\beta} = \partial^2\Theta/\partial w^{\dot\alpha}\partial w^{\dot\beta}$, the divergence coming from the $J \cdot \bar\partial^{-1} J \propto [w\, y]$ term can be renormalized away without affecting the metric.

This theorem can be proved using the same set of steps as its four-dimensional version Theorem 3.2. A slightly more streamlined approach based on [10] is described in the original work [8] where much of this material was first presented. This is reviewed in the rest of this section. The two cases need to be worked out separately.

Let's first outline the proof in the Kähler scalar case. When $T = 0$, a calculation shows that $\Sigma = \mathrm{d}\vartheta$ where $\vartheta = [\mu\,\mathrm{d}\mu] + 2h$ as in the previous section. Therefore $\mathrm{d}\Sigma = 0$, whence Σ is again conserved along the holomorphic curves $\mu^{\dot\alpha} = F^{\dot\alpha}(y,\lambda)$.

Let $p : M \times \mathbb{P}^1 \to \mathbb{P}\mathscr{T}$ denote the incidence relations map $(y, \lambda_\alpha) \mapsto (F^{\dot\alpha}, \lambda_\alpha)$. Then we find

$$p^*\Sigma = p^*(\theta^{\dot\alpha} \wedge \theta_{\dot\alpha}) = \lambda_\alpha\lambda_\beta\Sigma^{\alpha\beta}(y) \quad \mod \mathrm{D}\lambda\,. \tag{4.52}$$

Pulling back $\theta^{\dot\alpha} = \mathrm{d}\mu^{\dot\alpha} - \mathcal{L}_{\partial^{\dot\alpha}} h$ by p yields

$$(\mathrm{d}F^{\dot\alpha} - p^*\mathcal{L}_{\partial^{\dot\alpha}} h) \wedge (\mathrm{d}F_{\dot\alpha} + p^*\mathcal{L}_{\partial_{\dot\alpha}} h) = \lambda_\alpha\lambda_\beta\Sigma^{\alpha\beta} \quad \mod \mathrm{D}\lambda\,. \tag{4.53}$$

We will drop the p^* in what follows to keep the notation under control. To obtain the various heavenly descriptions, we take $h = \check\omega^2\varphi$ for some smooth φ. The calculation may be divided into the two cases $\check\omega = \lambda_1\lambda_2, \lambda_2^2$.

Keep in mind our conventions that $\lambda_\alpha = \lambda_1\kappa_{1\alpha} + \lambda_2\kappa_{2\alpha}$ and $\langle 1\,2\rangle = -1$. In the Kähler case $\check\omega = \lambda_1\lambda_2$, we can set $\lambda = \kappa_1$ or $\lambda = \kappa_2$ to find

$$\begin{aligned} \Sigma^{11} &\equiv \kappa_{1\alpha}\kappa_{1\beta}\Sigma^{\alpha\beta} = \mathrm{d}y^{\dot\alpha} \wedge \mathrm{d}y_{\dot\alpha}\,, \\ \Sigma^{22} &\equiv \kappa_{2\alpha}\kappa_{2\beta}\Sigma^{\alpha\beta} = \mathrm{d}\tilde y^{\dot\alpha} \wedge \mathrm{d}\tilde y_{\dot\alpha}\,. \end{aligned} \tag{4.54}$$

On the other hand, $\Sigma^{12} \equiv \kappa_{1\alpha}\kappa_{2\beta}\Sigma^{\alpha\beta}$ can be obtained in the two supposedly equivalent ways

$$\Sigma^{12} = \frac{1}{2}\left(\mathcal{L}_{\frac{\partial}{\partial\lambda_2}}\Sigma\right)\bigg|_{\lambda=\kappa_1} = \frac{1}{2}\left(\mathcal{L}_{\frac{\partial}{\partial\lambda_1}}\Sigma\right)\bigg|_{\lambda=\kappa_2}\,. \tag{4.55}$$

Define the space-time spinor quantities

$$\Omega^{\dot\alpha}(y, \tilde y) = \frac{\partial F^{\dot\alpha}}{\partial\lambda_2}\bigg|_{\lambda=\kappa_1} = \tilde y^{\dot\alpha} + \mathrm{m}^{\dot\alpha}(\kappa_1)\,, \tag{4.56}$$

$$\tilde\Omega^{\dot\alpha}(y, \tilde y) = \frac{\partial F^{\dot\alpha}}{\partial\lambda_1}\bigg|_{\lambda=\kappa_2} = y^{\dot\alpha} + \mathrm{m}^{\dot\alpha}(\kappa_2)\,. \tag{4.57}$$

having used $F^{\dot\alpha} = \lambda_1 y^{\dot\alpha} + \lambda_2\tilde y^{\dot\alpha} + \lambda_1\lambda_2\mathrm{m}^{\dot\alpha}$ to obtain the second equality in each of these. Then equating the two expressions for Σ^{12}, one obtains the constraint

$$\Sigma^{12} = \mathrm{d}y^{\dot\alpha} \wedge \mathrm{d}\Omega_{\dot\alpha} = \mathrm{d}\tilde y^{\dot\alpha} \wedge \mathrm{d}\tilde\Omega_{\dot\alpha}\,, \tag{4.58}$$

where the explicit dependence of Σ on h drops out due to the second order zeroes of h at $\lambda = \kappa_1, \kappa_2$. This is locally solved in terms of a scalar potential Ω by

$$\Omega_{\dot\alpha} = -\frac{\partial\Omega}{\partial y^{\dot\alpha}}\,, \qquad \tilde\Omega_{\dot\alpha} = \frac{\partial\Omega}{\partial\tilde y^{\dot\alpha}}\,. \tag{4.59}$$

This is our Kähler potential, because the resulting ASD 2-form Σ^{12} reads

$$\Sigma^{12} = -\Omega_{\dot\alpha\dot\beta}\,\mathrm{d}y^{\dot\alpha} \wedge \mathrm{d}\tilde y^{\dot\beta}\,. \tag{4.60}$$

As explained in Eq. (4.26), these values for $\Sigma^{\alpha\beta}$ come from a Kähler metric on our $4k$-dimensional manifold M. The constraint of Ricci-flatness also emerges automatically from the $\mathrm{d}\tilde{y}^{\dot\alpha} \wedge \mathrm{d}\tilde{y}_{\dot\alpha}$ component of the condition

$$\Sigma^{22} = \Sigma\Big|_{\lambda=\kappa_2} = \frac{1}{2}\left(\mathcal{L}^2_{\frac{\partial}{\partial\lambda_2}}\Sigma\right)\Big|_{\lambda=\kappa_1} . \tag{4.61}$$

One may easily verify that this yields Plebanski's first heavenly equation (the factors of $\mathcal{L}_{\partial_{\dot\alpha}} h$ again drop out in these combinations of the derivatives because of the second order zeroes of h).

To obtain the Plebanski scalar description, one needs to use the boundary conditions (4.40). Evaluating $p^*\Sigma$ at $\lambda = \kappa_1$ immediately yields $\Sigma^{11} = \mathrm{d}y^{\dot\alpha} \wedge \mathrm{d}y_{\dot\alpha}$. Similarly,

$$\Sigma^{12} = \frac{1}{2}\left(\mathcal{L}_{\frac{\partial}{\partial\lambda_2}}\Sigma\right)\Big|_{\lambda=\kappa_1} = \mathrm{d}y^{\dot\alpha} \wedge \mathrm{d}\left(\frac{\partial F_{\dot\alpha}}{\partial\lambda_2}\Big|_{\lambda=\kappa_1}\right) = \mathrm{d}y^{\dot\alpha} \wedge \mathrm{d}w_{\dot\alpha} \tag{4.62}$$

having now used the fourth order zero of h at $\lambda = \kappa_1$. Since h no longer vanishes at $\lambda = \kappa_2$, we observe that $\Sigma^{22} \neq \mathrm{d}w^{\dot\alpha} \wedge \mathrm{d}w_{\dot\alpha}$. Instead, it is easier to evaluate Σ^{22} via the formula

$$\Sigma^{22} = \frac{1}{2}\left(\mathcal{L}^2_{\frac{\partial}{\partial\lambda_2}}\Sigma\right)\Big|_{\lambda=\kappa_1} = \mathrm{d}w^{\dot\alpha} \wedge \mathrm{d}w_{\dot\alpha} + 2\,\mathrm{d}\Theta^{\dot\alpha} \wedge \mathrm{d}y_{\dot\alpha} , \tag{4.63}$$

where we have defined the space-time quantities

$$\Theta^{\dot\alpha}(y, w) = \frac{1}{2}\frac{\partial^2 F^{\dot\alpha}}{\partial\lambda_2^2}\Big|_{\lambda=\kappa_1} = \mathfrak{m}^{\dot\alpha}(\kappa_1) , \tag{4.64}$$

having used $F^{\dot\alpha} = \lambda_1 y^{\dot\alpha} + \lambda_2 w^{\dot\alpha} + \lambda_2^2 \mathfrak{m}^{\dot\alpha}$ to obtain the second equality. Since $\Sigma = \lambda_\alpha\lambda_\beta\Sigma^{\alpha\beta}$ is quadratic in λ_1, λ_2, the third derivative $\mathcal{L}^3_{\partial/\partial\lambda_2}\Sigma$ at $\lambda = \kappa_1$ must vanish. This yields the constraints

$$\frac{\partial\Theta_{\dot\alpha}}{\partial w^{\dot\beta}} = \frac{\partial\Theta_{\dot\beta}}{\partial w^{\dot\alpha}} , \qquad \frac{\partial\Theta_{\dot\alpha}}{\partial y^{\dot\beta}} = \frac{\partial}{\partial w^{\dot\alpha}}\left(\frac{1}{3!}\frac{\partial^3 F_{\dot\beta}}{\partial\lambda_2^3}\Big|_{\lambda=\kappa_1}\right) . \tag{4.65}$$

The first of these is solved in terms of a Plebanski scalar Θ by setting

$$\Theta_{\dot\alpha} = \frac{\partial\Theta}{\partial w^{\dot\alpha}} . \tag{4.66}$$

Using these along with the $\mathrm{d}w^{\dot\alpha} \wedge \mathrm{d}w^{\dot\beta}$ component of the vanishing of the fourth derivative $\mathcal{L}^4_{\partial/\partial\lambda_2}\Sigma$ at $\lambda = \kappa_1$, one obtains the second heavenly equation on Θ.

Equations (4.57) and (4.64) are all the data we need to prove Theorem 4.1. The equation of motion of the sigma model (4.49) is again the PDE for holomorphic curves,

$$\bar{\partial}_0 m_{\dot{\alpha}} = -\bar{\partial}_0 \lrcorner \mathcal{L}_{\partial_{\dot{\alpha}}} h \,. \tag{4.67}$$

Differentiating the corresponding on-shell action, we first obtain

$$\frac{\partial}{\partial y^{\alpha\dot{\alpha}}} S_h[m]\Big|_{\text{on-shell}} = \frac{1}{2\pi \mathrm{i}} \int \frac{\mathrm{D}\lambda}{\check{\omega}^2} \wedge \left\{ \frac{1}{2} \frac{\partial m^{\dot{\beta}}}{\partial y^{\alpha\dot{\alpha}}} \bar{\partial} m_{\dot{\beta}} + \frac{1}{2} m^{\dot{\beta}} \bar{\partial} \frac{\partial m_{\dot{\beta}}}{\partial y^{\alpha\dot{\alpha}}} + \mathcal{L}_{\partial/\partial y^{\alpha\dot{\alpha}}} h \right\} . \tag{4.68}$$

It is at this stage that working with the symplectic potential $\vartheta = [\mu\, \mathrm{d}\mu] + 2h$ turns out to be really efficient. This can be shown to obey the structure equation $\mathrm{d}\vartheta = \Sigma + 2T \mod \mathrm{D}\lambda$ by a calculation identical to the four-dimensional one (4.4). It follows that

$$\mathrm{d}h = \frac{1}{2} (\mathrm{d}\vartheta - \mathrm{d}\mu^{\dot{\alpha}} \wedge \mathrm{d}\mu_{\dot{\alpha}}) = \frac{1}{2} (\Sigma + 2T) - \frac{1}{2} \mathrm{d}\mu^{\dot{\alpha}} \wedge \mathrm{d}\mu_{\dot{\alpha}} \qquad \mod \mathrm{D}\lambda \,. \tag{4.69}$$

The $\bar{e}^0$ component of the variation of h in $y^{\alpha\dot{\alpha}}$ is then computed using Cartan's formula:

$$\begin{aligned}
\bar{\partial}_0|_y \lrcorner \mathcal{L}_{\partial/\partial y^{\alpha\dot{\alpha}}} h &= \bar{\partial}_0|_y \lrcorner \left\{ \frac{\partial}{\partial y^{\alpha\dot{\alpha}}} \lrcorner \mathrm{d}h + \mathrm{d}\left(\frac{\partial}{\partial y^{\alpha\dot{\alpha}}} \lrcorner h \right) \right\} \\
&= \frac{1}{2} \bar{\partial}_0|_y \lrcorner \frac{\partial}{\partial y^{\alpha\dot{\alpha}}} \lrcorner (\Sigma + 2T) - \frac{\partial F^{\dot{\beta}}}{\partial y^{\alpha\dot{\alpha}}} \bar{\partial}_0|_y F_{\dot{\beta}} + \bar{\partial}_0|_y \left(\frac{\partial}{\partial y^{\alpha\dot{\alpha}}} \lrcorner h \right) \\
&= -\frac{\partial F^{\dot{\beta}}}{\partial y^{\alpha\dot{\alpha}}} \bar{\partial}_0|_y F_{\dot{\beta}} + \bar{\partial}_0|_y \left(\frac{\partial}{\partial y^{\alpha\dot{\alpha}}} \lrcorner h \right) ,
\end{aligned} \tag{4.70}$$

where we have written $\bar{\partial}_0|_y$ for $\bar{\partial}_0$ at fixed $y^{\alpha\dot{\alpha}}$ to avoid any confusion. In the third line, we dropped the first term of the second line because $\bar{\partial}_0|_y \lrcorner \Sigma = 0$ along the holomorphic curves, and $T = 0$ by integrability of the complex structure deformation.

Feeding this Lie derivative back into (4.68), and dropping the $\bar{\partial}_0$-exact term using the boundary condition $h = \check{\omega}^2 \varphi$, we are left with

$$\frac{\partial}{\partial y^{\alpha\dot{\alpha}}} S_h[m]\Big|_{\text{on-shell}} = \frac{1}{2\pi \mathrm{i}} \int \frac{\mathrm{D}\lambda}{\check{\omega}^2} \wedge \left\{ \frac{1}{2} \frac{\partial m^{\dot{\beta}}}{\partial y^{\alpha\dot{\alpha}}} \bar{\partial} m_{\dot{\beta}} + \frac{1}{2} m^{\dot{\beta}} \bar{\partial} \frac{\partial m_{\dot{\beta}}}{\partial y^{\alpha\dot{\alpha}}} - \frac{\partial F^{\dot{\beta}}}{\partial y^{\alpha\dot{\alpha}}} \bar{\partial} F_{\dot{\beta}} \right\} . \tag{4.71}$$

Plugging in the equation of motion and the expansion $F^{\dot{\beta}} = y^{\beta\dot{\beta}} \lambda_\beta + m^{\dot{\beta}}$, we can kill terms by converting them into total $\bar{\partial}_0$ derivatives. This yields

$$\frac{\partial}{\partial y^{\alpha\dot{\alpha}}} S_h[m]\Big|_{\text{on-shell}} = -\frac{1}{2\pi \mathrm{i}} \int \frac{\mathrm{D}\lambda}{\check{\omega}^2} \wedge \lambda_\alpha \bar{\partial} m_{\dot{\alpha}}$$
$$= -\frac{1}{2\pi \mathrm{i}} \int \frac{\mathrm{D}\lambda}{\check{\omega}} \wedge \lambda_\alpha \bar{\partial} \mathfrak{m}_{\dot{\alpha}}\,, \tag{4.72}$$

having used $m^{\dot{\alpha}} = \check{\omega}\,\mathfrak{m}^{\dot{\alpha}}$ to obtain the second line.

In the Kähler case, the choice $\check{\omega} = \lambda_1\lambda_2$ allows us to localize the integral to $\lambda = \kappa_2$ by a further integration by parts. In particular, this yields the $\partial/\partial\tilde{y}^{\dot{\alpha}} = \partial/\partial y^{2\dot{\alpha}}$ derivative

$$\frac{\partial}{\partial \tilde{y}^{\dot{\alpha}}} S_h[m]\Big|_{\text{on-shell}} = -\mathfrak{m}_{\dot{\alpha}}(\kappa_2) = y_{\dot{\alpha}} - \frac{\partial\Omega}{\partial\tilde{y}^{\dot{\alpha}}}\,, \tag{4.73}$$

having invoked (4.57). This allows us to identify Ω with the right hand side of (4.50). Similarly, in the Plebanski scalar description, we set $\check{\omega} = \lambda_2^2$. This localizes the remaining integral to $\lambda = \kappa_1$. In this case, we may invoke (4.64) to obtain

$$\frac{\partial}{\partial w^{\dot{\alpha}}} S_h[m]\Big|_{\text{on-shell}} = -\mathfrak{m}^{\dot{\alpha}}(\kappa_1) = -\frac{\partial\Theta}{\partial w^{\dot{\alpha}}}\,. \tag{4.74}$$

This establishes (4.51), up to the irrelevant ambiguity of shifting Θ by functions linear in $w^{\dot{\alpha}}$.

4.3 Turning on a Cosmological Constant

Up till now, we have focused our attention on hyperkähler manifolds. These were in particular Ricci-flat. For many geometric applications, it is also useful to study quaternion-Kähler manifolds [11]. This is the case of $\mathrm{Sp}(1)\cdot\mathrm{Sp}(k) \equiv (\mathrm{Sp}(1)\times\mathrm{Sp}(k))/\mathbb{Z}_2$ holonomy. It corresponds to $4k$-real-dimensional Riemannian manifolds (M, g) whose Riemann tensor 2-form lies in the Lie algebra $\mathfrak{sp}(1)\oplus\mathfrak{sp}(k)$,

$$R_{\alpha\dot{\alpha}\beta\dot{\beta}} = \epsilon_{\dot{\alpha}\dot{\beta}}\, R_{\alpha\beta} + \epsilon_{\alpha\beta}\, \tilde{R}_{\dot{\alpha}\dot{\beta}}\,. \tag{4.75}$$

Such a manifold is automatically Einstein, i.e., has Ricci tensor proportional to the metric. This happens to be equivalent to the condition that the $\mathfrak{sp}(1)$-valued Riemann curvature $R_{\alpha\beta}$ be proportional to the ASD 2-forms $\Sigma^{\alpha\beta}$:

$$R^{\alpha\beta} = \Lambda\,\Sigma^{\alpha\beta}\,. \tag{4.76}$$

The constant of proportionality Λ is a 4k-dimensional avatar of the cosmological constant familiar from GR. In the special case of $k = 1$, quaternion-Kähler manifolds are the same as self-dual Einstein manifolds. When $\Lambda = 0$, we get back the hyperkähler case.

In the rest of this chapter, we outline the construction of a twistor sigma model that packages the data of quaternion-Kähler metrics into a scalar known as the *Swann scalar*. This will act as the Kähler scalar of a hyperkähler structure on the twistor spaces of quaternion-Kähler manifolds [12]. On M, this will reduce to a scalar known as the *Przanowski scalar* [13]. For more details of the associated geometry as well as references to the literature, the reader is referred to [8].

Let us start by describing the construction of twistor spaces for such manifolds. The generators of the $\mathfrak{sp}(1)$ factor of the holonomy algebra $\mathfrak{sp}(1) \oplus \mathfrak{sp}(k)$ provide a triplet of almost complex structures I, J, K on M. Because $\mathfrak{sp}(1) = \mathfrak{su}(2)$, they can be chosen to satisfy the quaternion algebra. By analogy with the hyperkähler case, the twistor space of (M, g) is defined to be the 2-sphere bundle $S^2 \to \mathbb{P}\mathscr{T} \to M$ whose local sections are almost complex structures of the form $aI + bJ + cK$ with $a^2 + b^2 + c^2 = 1$. It again admits an integrable almost complex structure. At a point $x \in M$ and $(a, b, c) \in S^2$ on the fiber over M, this almost complex structure is defined to be the direct sum of the complex structure of $\mathbb{P}^1$ with the endomorphism $aI + bJ + cK$ evaluated at x. Unfortunately, $\mathbb{P}\mathscr{T}$ no longer fibers over $\mathbb{P}^1$ in any canonical way. This is a reflection of the fact that the quaternion-Kähler manifold M may not admit any globally defined complex structure, even though it admits local almost complex structures $aI + bJ + cK$ with respect to which the metric g is locally hermitian.

A useful construction of quaternion-Kähler twistor spaces is found by again deforming the flat twistor space $\mathbb{PT} = \mathbb{P}^{2k+1} - \mathbb{P}^{2k-1}$. Let $Z^A = (\mu^{\dot\alpha}, \lambda_\alpha)$, $\alpha = 1, 2$, $\dot\alpha = \dot 1, \ldots, \dot{2k}$, be homogeneous coordinates on $\mathbb{P}^{2k+1}$. In this case, instead of studying a fiberwise symplectic structure on the fibers of $\mathbb{PT} = \mathbb{C}^{2k} \otimes \mathcal{O}(1)$, one defines a weighted symplectic structure that encodes the cosmological constant:

$$\{f, g\} = I^{AB}\, \mathcal{L}_{\partial_A} f \wedge \mathcal{L}_{\partial_B} g\,, \qquad f, g \in \Omega^{0,p}(\mathbb{PT})\,,$$

where $\partial_A = \partial/\partial Z^A$. Here and in what follows, I^{AB} and I_{AB} denote certain *infinity twistor* skew forms preserved by $\mathrm{Sp}(k) \times \mathrm{Sp}(1)$:

$$I^{AB} = \begin{pmatrix} \epsilon^{\dot\alpha\dot\beta} & 0 \\ 0 & \Lambda\epsilon_{\alpha\beta} \end{pmatrix}, \qquad I_{AB} = \begin{pmatrix} \Lambda\epsilon_{\dot\alpha\dot\beta} & 0 \\ 0 & \epsilon^{\alpha\beta} \end{pmatrix}, \qquad I^{AB} I_{AC} = \Lambda \delta^A_C\,,$$

with $\epsilon_{\alpha\beta}$, $\epsilon^{\alpha\beta}$ and $\epsilon_{\dot\alpha\dot\beta}$, $\epsilon^{\dot\alpha\dot\beta}$ being the $\mathrm{Sp}(1)$ and $\mathrm{Sp}(k)$ invariant skew forms respectively. As $\Lambda \to 0$, the $\mathrm{Sp}(1)$ decouples and the symplectic structure reduces to the fiberwise Poisson bracket $\{f, g\} = \epsilon^{\dot\alpha\dot\beta} \mathcal{L}_{\partial_{\dot\alpha}} f \wedge \mathcal{L}_{\partial_{\dot\beta}} g$ encountered in the hyperkähler case.

Twistor spaces that yield new quaternion-Kähler geometries are found by a turning on a hamiltonian complex structure deformation

$$\bar\partial \mapsto \bar\partial + I^{AB}\, \mathcal{L}_{\partial_A} h\, \partial_B\,, \qquad h \equiv h_A \mathrm{d}\hat Z^A \in \mathcal{O}(2)\,. \tag{4.77}$$

Here, the background Dolbeault operator is $\bar{\partial} = \mathrm{d}\hat{Z}^A\,\hat{\partial}_A \equiv \mathrm{d}\hat{Z}^A\,\partial/\partial\hat{Z}^A$, written in terms of quaternionic conjugates $\hat{Z}^A := (\hat{\mu}^{\dot{\alpha}}, \hat{\lambda}_\alpha)$. The hamiltonian h is subject to the constraints $Z^A\partial_A h = 2h$ and $\hat{Z}^A h_A = 0$ in order for it to define a (0, 1)-form of weight 2 on $\mathbb{P}^{2k+1}$. The deformed almost complex structure is integrable if h satisfies

$$T := \bar{\partial} h + \frac{1}{2}\{h, h\} = 0\,. \tag{4.78}$$

The $(1, 0)$-forms of the deformed complex manifold $\mathbb{P}\mathscr{T}$ read

$$\theta^A = \mathrm{d}Z^A + I^{AB}\mathcal{L}_{\partial_B} h\,. \tag{4.79}$$

A twistor curve $Y : Z^A = Z^A(y, \sigma)$ labeled by moduli y is defined to be a holomorphic curve of degree 1 in $\sigma_\alpha \in \mathbb{P}^1$, with normal bundle $\mathbb{C}^{2k} \otimes \mathcal{O}(1)$, that satisfies the Cauchy-Riemann equations

$$\bar{\partial}_0 \lrcorner \left(\theta^A|_Y\right) = 0 \implies \bar{\partial}_0 Z^A = \bar{\partial}_0 \lrcorner \left(I^{BA}\mathcal{L}_{\partial_B} h|_Y\right), \tag{4.80}$$

where $\bar{\partial}_0 = \langle\hat{\sigma}\,\sigma\rangle\,\sigma \cdot \partial_{\hat{\sigma}}$ now denotes the antiholomorphic vector field on $\mathbb{P}^1$ in some abstract homogeneous coordinates σ_α that no longer equal λ_α. If the deformation is small enough, such curves continue to foliate $\mathbb{P}\mathscr{T}$ and are in one-to-one correspondence with points of a deformed $4k$-dimensional manifold M, on which we will shortly construct a quaternion-Kähler structure. Denoting by $p : M \times \mathbb{P}^1 \to \mathbb{P}\mathscr{T}$ the map $(y, \sigma) \mapsto Z^A = Z^A(y, \sigma)$, we find a 2-sphere bundle structure $\mathbb{P}\mathscr{T} \to M$ as needed.

Let us abbreviate contractions of twistor indices by

$$\langle Z_1, Z_2\rangle \equiv I_{AB} Z_1^B Z_2^A\,, \tag{4.81}$$

with the placement of indices mimicking the angle brackets $\langle\lambda\,\kappa\rangle = \lambda^\beta\kappa_\beta = \epsilon_{\alpha\beta}\lambda^\beta\kappa^\alpha$. To obtain the deformed quaternion-Kähler metric, one endows $\mathbb{P}\mathscr{T}$ with a weighted holomorphic contact structure

$$\tau := \langle Z, \theta\rangle = \langle Z, \mathrm{d}Z\rangle + 2\Lambda h\,. \tag{4.82}$$

In obtaining the second equality, we have used the homogeneity constraint $Z^A\partial_A h = 2h$ and the infinity twistor contraction $I_{AB}I^{AC} = \delta_B^C$. The first step in obtaining the quaternion-Kähler metric is to show that $p^*\tau$ is holomorphic along the twistor curves.

To do this, we begin by calculating its exterior derivative

$$\begin{aligned}\mathrm{d}\tau &= I_{AB}\,\mathrm{d}Z^B \wedge \mathrm{d}Z^A + 2\Lambda\,\mathrm{d}Z^A \wedge \mathcal{L}_{\partial_A} h + 2\Lambda\bar{\partial} h\\ &= I_{AB}\,(\theta^B - I^{BC}\mathcal{L}_{\partial_C} h) \wedge (\theta^A - I^{AD}\mathcal{L}_{\partial_D} h) + 2\Lambda\,(\theta^A - I^{AB}\mathcal{L}_{\partial_B} h) \wedge \mathcal{L}_{\partial_A} h + 2\Lambda\bar{\partial} h\end{aligned} \tag{4.83}$$

which condenses to the structure equation

$$d\tau = \Lambda(\Sigma + 2T) \tag{4.84}$$

in terms of the 2-forms

$$\Sigma = \frac{1}{\Lambda} I_{AB}\, \theta^B \wedge \theta^A . \tag{4.85}$$

These 2-forms generalize the Gindikin 2-form $\Sigma = \theta^{\dot\alpha} \wedge \theta_{\dot\alpha}$ from the hyperkähler case. From (4.84), we can calculate the variation of τ along a twistor curve:

$$\mathcal{L}_{\bar\partial_0} p^*\tau = \bar\partial_0 \lrcorner\, p^* d\tau + d(\bar\partial_0 \lrcorner\, p^*\tau) . \tag{4.86}$$

Since τ is a $(1, 0)$-form, it has no antiholomorphic component along holomorphic curves, so the second term vanishes. Similarly, since Σ is a $(2, 0)$-form, the first term can be simplified using (4.84):

$$\mathcal{L}_{\bar\partial_0} p^*\tau = 2\Lambda\, \bar\partial_0 \lrcorner\, p^* T . \tag{4.87}$$

Hence, τ is global along the curves on the support of $T = 0$.

To extract the space-time content of τ, we need to make $p^*\tau$ explicit. To this end, let us parametrize our twistor curves explicitly. Since a twistor curve has normal bundle $\mathbb{C}^{2k} \otimes \mathcal{O}(1)$, it has $4k$ complex moduli. Adding to this the $\mathrm{GL}(2, \mathbb{C})$ freedom in reparametrizations of σ_α, we obtain $4k + 4$ degrees of freedom per curve. If we demand that our curve passes through two points $\mathcal{Z}^A$ and $\tilde{\mathcal{Z}}^A$ in $\mathbb{P}\mathscr{T}$, each containing $2k + 2$ components, then we get enough equations to completely fix the curve. Hence, we can parametrize each curve by a pair of points $\mathcal{Z}^A$, $\tilde{\mathcal{Z}}^A$ acting as moduli. We write such a curve as

$$Z^A = F^A(\mathcal{Z}, \tilde{\mathcal{Z}}, \sigma) , \tag{4.88}$$

where the data of points $y \in M$ is understood as being equivalent to the data of $\mathcal{Z}^A, \tilde{\mathcal{Z}}^A$, up to the induced action of $\mathrm{GL}(2, \mathbb{C})$ killing 4 of the $4k + 4$ degrees of freedom. For example, in flat space, $\mathcal{Z}^A = (y^{\dot\alpha}, \kappa_{1\alpha})$ and $\tilde{\mathcal{Z}}^A = (\tilde{y}^{\dot\alpha}, \kappa_{2\alpha})$.

Let $\sigma_\alpha = \sigma_1\kappa_{1\alpha} + \sigma_2\kappa_{2\alpha}$. Fix the $\mathrm{GL}(2, \mathbb{C})$ freedom in the definition of σ_α, as well as the curve moduli, by imposing the boundary conditions

$$\mathcal{Z}^A = F^A(\mathcal{Z}, \tilde{\mathcal{Z}}, \kappa_1) , \qquad \tilde{\mathcal{Z}}^A = F^A(\mathcal{Z}, \tilde{\mathcal{Z}}, \kappa_2) . \tag{4.89}$$

Since τ is holomorphic along each such curve, we start by demanding that $p^*\tau$ restrict to the standard holomorphic 1-form $\sigma^\alpha d\sigma_\alpha$ at fixed $\mathcal{Z}$, $\tilde{\mathcal{Z}}$ (i.e., at fixed $y \in M$), up to a constant of proportionality $\Omega(\mathcal{Z}, \tilde{\mathcal{Z}})$:

$$p^*\tau|_{\mathcal{Z},\tilde{\mathcal{Z}}} \overset{!}{=} \Omega(\mathcal{Z}, \tilde{\mathcal{Z}})\, \langle \sigma\, d\sigma \rangle . \tag{4.90}$$

Here, we have taken Ω to be independent of σ_α by invoking Liouville's theorem, because Ω must necessarily be a weight 0 globally holomorphic function in σ_α.

In the four-dimensional case, the extraction of the data of a spacetime metric from the globality of τ has been discussed in [14], and it is easily generalized to higher dimensions. Recall that $\tau = \langle Z, \mathrm{d}Z\rangle + 2\Lambda h$ and (locally if needed) choose a gauge in which $p^*h = \sigma_1^2\sigma_2^2\varphi$ with φ smooth at $\sigma = \kappa_1, \kappa_2$ (just as with the hyperkähler case). Evaluating $p^*\tau$ yields

$$p^*\tau = \langle F, \mathrm{d}F\rangle + 2\Lambda\, p^*h\,. \tag{4.91}$$

Imposing (4.90) on this and setting $\sigma = \kappa_1, \kappa_2$ generates the constraints

$$\Omega = I_{AB}\mathcal{Z}^B\Omega^A = I_{AB}\tilde{\mathcal{Z}}^B\tilde{\Omega}^A\,, \tag{4.92}$$

where we have introduced the quantities

$$\Omega^A(\mathcal{Z},\tilde{\mathcal{Z}}) = -\frac{\partial F^A}{\partial\sigma_2}\bigg|_{\sigma=\kappa_1}\,,\qquad \tilde{\Omega}^A(\mathcal{Z},\tilde{\mathcal{Z}}) = \frac{\partial F^A}{\partial\sigma_1}\bigg|_{\sigma=\kappa_2}\,. \tag{4.93}$$

Next, we use the holomorphicity of the components of $p^*\tau$ pointing along M.

Since τ has weight 2 in σ, the pullback $p^*\tau$ at fixed σ takes the form

$$p^*\tau|_\sigma = \Omega\sigma^\alpha\sigma^\beta\Gamma_{\alpha\beta} = \Omega\left(\sigma_1^2\Gamma_{22} - 2\sigma_1\sigma_2\Gamma_{12} + \sigma_2^2\Gamma_{11}\right)\,, \tag{4.94}$$

where $\Gamma_{\alpha\beta}$ are symmetric 1-forms spanned by $\mathrm{d}\mathcal{Z}^a, \mathrm{d}\tilde{\mathcal{Z}}^A$.[1] From this form of the pullback, we deduce the three equations

$$\Omega\Gamma_{12} = -\frac{1}{2}\mathcal{L}_{\partial_{\sigma_2}}p^*\tau\big|_{\sigma=\kappa_1} = -\frac{1}{2}\mathcal{L}_{\partial_{\sigma_1}}p^*\tau\big|_{\sigma=\kappa_2} \tag{4.95}$$

$$\Omega\Gamma_{22} = p^*\tau\big|_{\sigma=\kappa_1} = \frac{1}{2}\mathcal{L}^2_{\partial_{\sigma_1}}p^*\tau\big|_{\sigma=\kappa_2} \tag{4.96}$$

$$\Omega\Gamma_{11} = p^*\tau\big|_{\sigma=\kappa_2} = \frac{1}{2}\mathcal{L}^2_{\partial_{\sigma_2}}p^*\tau\big|_{\sigma=\kappa_1}\,. \tag{4.97}$$

The second equality in each of these equations leads to the following family of constraints on the incidence relations:

$$\langle\mathcal{Z},\mathrm{d}\Omega\rangle + \langle\Omega,\mathrm{d}\mathcal{Z}\rangle + \langle\tilde{\mathcal{Z}},\mathrm{d}\tilde{\Omega}\rangle + \langle\tilde{\Omega},\mathrm{d}\tilde{\mathcal{Z}}\rangle = 0 \tag{4.98}$$

$$\langle\tilde{\Omega},\mathrm{d}\tilde{\Omega}\rangle + \langle\tilde{\mathcal{Z}},\mathrm{d}\tilde{N}\rangle + \langle\tilde{N},\mathrm{d}\tilde{\mathcal{Z}}\rangle = \langle\mathcal{Z},\mathrm{d}\mathcal{Z}\rangle \tag{4.99}$$

$$\langle\Omega,\mathrm{d}\Omega\rangle + \langle\mathcal{Z},\mathrm{d}N\rangle + \langle N,\mathrm{d}\mathcal{Z}\rangle = \langle\tilde{\mathcal{Z}},\mathrm{d}\tilde{\mathcal{Z}}\rangle\,, \tag{4.100}$$

[1] They are secretly the spin connection coefficients of the quaternion-Kähler metric on M, but we will avoid deriving this here. See [14] for details.

having further defined $N^A = \frac{1}{2}\partial^2_{\sigma_1} F^A|_{\sigma=\kappa_2}$ and $\tilde{N}^A = \frac{1}{2}\partial^2_{\sigma_2} F^A|_{\sigma=\kappa_1}$. Taking an exterior derivative of (4.92), we can simplify the first of these down to

$$\mathrm{d}\Omega = \langle \mathrm{d}\mathcal{Z}, \Omega\rangle + \langle \mathrm{d}\tilde{\mathcal{Z}}, \tilde{\Omega}\rangle \implies \Omega^A = \frac{I^{AB}}{\Lambda}\frac{\partial\Omega}{\partial\mathcal{Z}^B}, \quad \tilde{\Omega}^A = \frac{I^{AB}}{\Lambda}\frac{\partial\Omega}{\partial\tilde{\mathcal{Z}}^B}. \tag{4.101}$$

Taking exterior derivatives of the second and third equations and comparing components of $\mathrm{d}\mathcal{Z}^A \wedge \mathrm{d}\mathcal{Z}^B, \mathrm{d}\tilde{\mathcal{Z}}^A \wedge \mathrm{d}\tilde{\mathcal{Z}}^B$ then produces PDEs for Ω:

$$I^{AB}\frac{\partial^2\Omega}{\partial\mathcal{Z}^A\partial\tilde{\mathcal{Z}}^C}\frac{\partial^2\Omega}{\partial\mathcal{Z}^B\partial\tilde{\mathcal{Z}}^D} = I_{CD} = I^{AB}\frac{\partial^2\Omega}{\partial\tilde{\mathcal{Z}}^A\partial\mathcal{Z}^C}\frac{\partial^2\Omega}{\partial\tilde{\mathcal{Z}}^B\partial\mathcal{Z}^D}. \tag{4.102}$$

The astute reader may immediately recognize these as the heavenly equations for some $(4k+4)$-real-dimensional hyperkähler geometry, but what is this geometry?

It is in fact a hyperkähler structure on *non-projective* twistor space $\mathscr{T}$. To define $\mathscr{T}$, one splits the tangent bundle of M as $TM = \mathbb{S}\otimes\tilde{\mathbb{S}}$, where $\mathbb{S}$ is a bundle of 2-spinors that transforms as a representation of Sp(1) and $\tilde{\mathbb{S}}$ is a bundle transforming in Sp(k) under parallel transport. Then $\mathscr{T}$ may be taken to be the total space of $\mathbb{S}$, and $\mathbb{P}\mathscr{T}$ is obtained by projectivizing the $\mathbb{C}^2$ fibers of $\mathscr{T}$. If we take Z^A as $2k+2$ complex coordinates on $\mathscr{T}$, then $\Omega(Z,\hat{Z})$ gives rise to a hyperkähler metric on $\mathscr{T}$:

$$g_{\mathscr{T}} = 2\frac{\partial^2\Omega}{\partial Z^A\partial\hat{Z}^B}\,\mathrm{d}Z^A\,\mathrm{d}\hat{Z}^B. \tag{4.103}$$

The first heavenly equation on $\Omega(Z,\hat{Z})$ is obtained from (4.102) by making the replacements $\mathcal{Z}^A \to Z^A$, $\tilde{\mathcal{Z}}^A \to \hat{Z}^A$. This hyperkähler structure was constructed by Swann in [12], hence the scalar Ω is known as a *Swann scalar*. $\mathscr{T}$ may thus be viewed as a hyperkähler cone over M, and is also known as the Swann bundle in the literature.

We can always locally use the GL$(2,\mathbb{C})$ freedom in the curves to set

$$\mathcal{Z}^A = (y^{\dot\alpha}, \kappa_{1\alpha}), \quad \tilde{\mathcal{Z}}^A = (\tilde{y}^{\dot\alpha}, \kappa_{2\alpha}). \tag{4.104}$$

The quaternion-Kähler metric on M obtained from Ω takes the form

$$g = 2\frac{\partial^2 K}{\partial y^{\dot\alpha}\partial\tilde{y}^{\dot\beta}}\,\mathrm{d}y^{\dot\alpha}\,\mathrm{d}\tilde{y}^{\dot\beta} + \frac{2}{\Lambda}\,\mathrm{e}^{2\Lambda K}\,[y\,\mathrm{d}y][\tilde{y}\,\mathrm{d}\tilde{y}]. \tag{4.105}$$

This form of the metric was first constructed in [13], so the scalar $K(y,\tilde{y})$ is known as the *Przanowski scalar*. It is defined in terms of Ω by

$$K(y,\tilde{y}) = -\frac{1}{\Lambda}\log\left(\frac{\Omega(\mathcal{Z},\tilde{\mathcal{Z}})}{\Lambda}\right). \tag{4.106}$$

We leave it as an exercise to check that the metric (4.105) satisfies the Einstein condition $R_{\alpha\beta} = \Lambda \Sigma_{\alpha\beta}$ in terms of a suitably defined triplet of ASD 2-forms $\Sigma^{\alpha\beta}$.

Finally, we return to the twistor sigma model. As we have reduced the construction of a quaternion-Kähler metric on M to that of a hyperkähler metric on $\mathscr{T}$, we can recycle the sigma models of the previous section. But now they also a simple geometric characterization in terms of the contact structure τ. Since we already used σ_α as the curve parameter in this section, let us introduce a different coordinate $\zeta_\alpha \in \mathbb{P}^1$ by the degree -1 map

$$\sigma_\alpha = \frac{\zeta_\alpha}{\zeta_1 \zeta_2} . \tag{4.107}$$

The twistor sigma model with a cosmological constant can be divided into two parts,

$$S_{h,\Lambda}[Z] = S_{\text{kinetic}} + S_{\text{source}} . \tag{4.108}$$

Here, S_{kinetic} is an action that governs degree -1 maps $\mathbb{P}^1 \to \mathbb{P}\mathscr{T}$ of the form $Z^A \mapsto Z^A(\zeta)$. It is given by integrating the pullback of τ by this map over $\mathbb{P}^1$:

$$S_{\text{kinetic}} = \frac{1}{4\pi \mathrm{i} \Lambda} \int_{\mathbb{P}^1} \mathrm{D}\zeta \wedge \tau = \frac{1}{2\pi \mathrm{i} \Lambda} \int_{\mathbb{P}^1} \mathrm{D}\zeta \wedge \left(\frac{1}{2} \langle Z, \bar{\partial} Z \rangle + \Lambda h(Z) \right) , \tag{4.109}$$

where $\mathrm{D}\zeta = \zeta^\alpha \mathrm{d}\zeta_\alpha$ and $\bar{\partial} = \mathrm{d}\bar{\zeta} \cdot \partial_{\bar{\zeta}}$. Because τ has degree 2 in Z^A, and Z^A has degree -1 in ζ_α, the $\mathbb{P}^1$ integral is well-defined. The second term is the source term

$$S_{\text{source}} = \frac{1}{\Lambda} \int_{\mathbb{P}^1} \mathrm{D}\zeta \wedge \left(\langle \mathcal{Z}, Z(\zeta) \rangle \, \bar{\delta}(\zeta_2) + \langle \tilde{\mathcal{Z}}, Z(\zeta) \rangle \, \bar{\delta}(\zeta_1) \right) . \tag{4.110}$$

One interesting feature ensuring backward compatibility is the fact that as $\Lambda \to 0$, these reduce to the kinetic and source terms of the twistor sigma model associated with the hyperkähler case. Indeed, one can use $\langle Z, \bar{\partial} Z \rangle = \langle \lambda \, \bar{\partial} \lambda \rangle + \Lambda [\mu \, \bar{\partial} \mu]$ to take the $\Lambda \to 0$ limit. In this limit, the saddle point approximation provided by the equation of motion of λ_α becomes exact. For the choice (4.104), the latter is given by $\bar{\partial} \lambda^\alpha = 2\pi \mathrm{i} \, \kappa_1^\alpha \bar{\delta}(\zeta_2) + 2\pi \mathrm{i} \, \kappa_2^\alpha \bar{\delta}(\zeta_1)$. This is solved by $\lambda_\alpha = \zeta_\alpha / \zeta_1 \zeta_2$, which allows us to identify $\lambda_\alpha = \sigma_\alpha$. The remaining effective action for $\mu^{\dot\alpha}$ is precisely our previous sigma model (4.44) for the Kähler case.

Finally, when $\Lambda \neq 0$, let's see how the on-shell action of this model reproduces the Swann scalar. Performing the field redefinition

$$Z^A(\zeta) = \frac{\mathcal{Z}^A}{\zeta_2} + \frac{\tilde{\mathcal{Z}}^A}{\zeta_1} + \mathfrak{m}^A(\zeta) \tag{4.111}$$

kills the sources and yields

$$S_{h,\Lambda}[Z] = \frac{1}{\Lambda} \langle \mathcal{Z}, \tilde{\mathcal{Z}} \rangle + \frac{1}{2\pi \mathrm{i} \Lambda} \int \mathrm{D}\sigma \wedge \left(\frac{1}{2} \langle \mathfrak{m}, \bar{\partial} \mathfrak{m} \rangle + \frac{\Lambda p^* h}{\sigma_1^2 \sigma_2^2} \right) , \tag{4.112}$$

having used $\mathrm{m}^A(\zeta) = \sigma_1\sigma_2\mathrm{m}^A(\sigma)$ to return to the coordinates $\sigma_\alpha = \zeta_\alpha/\zeta_1\zeta_2$. Here, $p^*h \equiv h(F)$ is understood as a function of $Z^A = F^A \equiv \sigma_1 \mathcal{Z}^A + \sigma_2\tilde{\mathcal{Z}}^A + \sigma_1\sigma_2\mathrm{m}^A(\sigma)$. It is easily confirmed that the equation of motion of m^A is the condition of holomorphicity (4.80).

If we take this action on-shell and differentiate with respect to $\mathcal{Z}^A$, we obtain

$$I^{AB}\frac{\partial}{\partial \mathcal{Z}^B} S_{h,\Lambda}[Z]\big|_{\text{on-shell}} = \tilde{\mathcal{Z}}^A + \frac{I^{AB}}{4\pi \mathrm{i}\Lambda}\int \mathrm{D}\sigma \wedge \left(\left\langle \frac{\partial \mathrm{m}}{\partial \mathcal{Z}^B}, \bar{\partial}\mathrm{m}\right\rangle + \left\langle \mathrm{m}, \bar{\partial}\frac{\partial \mathrm{m}}{\partial \mathcal{Z}^B}\right\rangle + \frac{2\Lambda}{\sigma_1^2\sigma_2^2}\mathcal{L}_{\partial/\partial \mathcal{Z}^B} p^*h\right). \tag{4.113}$$

Since $\tau = \langle Z, \mathrm{d}Z\rangle + 2\Lambda h$, the structure equation $\mathrm{d}\tau = \Lambda(\Sigma + 2T)$ obtained in (4.84) allows us to reverse engineer $\mathrm{d}h$:

$$\mathrm{d}h = \frac{1}{2\Lambda}(\mathrm{d}\tau - \langle \mathrm{d}Z, \mathrm{d}Z\rangle) = \frac{1}{2}(\Sigma + 2T) - \frac{\langle \mathrm{d}Z, \mathrm{d}Z\rangle}{2\Lambda}, \tag{4.114}$$

where $\langle \mathrm{d}Z, \mathrm{d}Z\rangle \equiv I_{AB}\mathrm{d}Z^B \wedge \mathrm{d}Z^A$. From this, we read off the Lie derivative

$$\begin{aligned}\mathcal{L}_{\partial/\partial \mathcal{Z}^B} p^*h &= \frac{\partial}{\partial \mathcal{Z}^B} \lrcorner\, p^*\mathrm{d}h + \mathrm{d}\left(\frac{\partial}{\partial \mathcal{Z}^B} \lrcorner\, p^*h\right) \\ &= \frac{1}{2}\frac{\partial}{\partial \mathcal{Z}^B} \lrcorner\, p^*(\Sigma + 2T) - \frac{1}{\Lambda}\left\langle \frac{\partial F}{\partial \mathcal{Z}^B}, \mathrm{d}F\right\rangle + \mathrm{d}\left(\frac{\partial}{\partial \mathcal{Z}^B} \lrcorner\, p^*h\right).\end{aligned} \tag{4.115}$$

When this is plugged into (4.113), the first term vanishes because $T = 0$ and because $p^*\Sigma$ has no antiholomorphic component along the holomorphic curves. The last term also drops out by virtue of being exact. Hence, keeping only the contribution of the middle term, we find

$$I^{AB}\frac{\partial}{\partial \mathcal{Z}^B} S_{h,\Lambda}[Z]\big|_{\text{on-shell}} = \tilde{\mathcal{Z}}^A + \frac{I^{AB}}{4\pi \mathrm{i}\Lambda}\int \mathrm{D}\sigma \wedge \left(\left\langle \frac{\partial \mathrm{m}}{\partial \mathcal{Z}^B}, \bar{\partial}\mathrm{m}\right\rangle + \left\langle \mathrm{m}, \bar{\partial}\frac{\partial \mathrm{m}}{\partial \mathcal{Z}^B}\right\rangle - \frac{2}{\sigma_1^2\sigma_2^2}\left\langle \frac{\partial F}{\partial \mathcal{Z}^B}, \bar{\partial}F\right\rangle\right). \tag{4.116}$$

Plugging in $F^A = \sigma_1\mathcal{Z}^A + \sigma_2\tilde{\mathcal{Z}}^A + \sigma_1\sigma_2\mathrm{m}^A$ and dropping boundary terms reduces this to

$$\begin{aligned}I^{AB}\frac{\partial}{\partial \mathcal{Z}^B} S_{h,\Lambda}[Z]\big|_{\text{on-shell}} &= \tilde{\mathcal{Z}}^A + \frac{I^{AB}}{4\pi \mathrm{i}\Lambda}\int \mathrm{D}\sigma \wedge \left\{\bar{\partial}\left\langle \mathrm{m}, \frac{\partial \mathrm{m}}{\partial \mathcal{Z}^B}\right\rangle - \frac{2}{\sigma_2} I_{CB}\bar{\partial}\mathrm{m}^C\right\} \\ &= \tilde{\mathcal{Z}}^A + \mathrm{m}^A(\kappa_1) \\ &= \frac{\partial F^A}{\partial \sigma_2}\bigg|_{\sigma = \kappa_1}.\end{aligned} \tag{4.117}$$

Recalling (4.93) and (4.101), we observe the sequence of equalities

$$I^{AB} \frac{\partial}{\partial \mathcal{Z}^B} S_{h,\Lambda}[Z]\Big|_{\text{on-shell}} = -\Omega^A = -\frac{I^{AB}}{\Lambda} \frac{\partial \Omega}{\partial \mathcal{Z}^B}, \tag{4.118}$$

from which we finally extract one of our main results

Theorem 4.2 *The on-shell action of the twistor sigma model* (4.108) *computes the Swann scalar of* $\mathbb{P}\mathscr{T}$:

$$\Omega(\mathcal{Z}, \tilde{\mathcal{Z}}) = -\Lambda\, S_{h,\Lambda}[Z]\Big|_{\text{on-shell}}. \tag{4.119}$$

This satisfies the heavenly equations (4.102) *for the hyperkähler structure on non-projective twistor space.*

This concludes our foray into geometric generalizations of our twistor sigma models. The $\Lambda \neq 0$ generalization is also expected to find applications to the study of scattering amplitudes in (A)dS$_4$, though we will focus on the $\Lambda = 0$ case for the rest of this thesis.

References

1. Frost H, Gürdogan O, Mason L (2023) Amplitudes at strong coupling as hyperkähler scalars. arXiv:2306.17044
2. Bittleston R, Sharma A, Skinner D (2022) Quantizing the non-linear graviton. arXiv:2208.12701
3. Geyer Y, Lipstein AE, Mason LJ (2014) Ambitwistor strings in four dimensions. Phys Rev Lett 113(8):081602. arXiv:1404.6219
4. Hitchin NJ, Karlhede A, Lindstrom U, Rocek M (1987) Hyperkahler Metrics and Supersymmetry. Commun Math Phys 108:535
5. Galperin A, Ivanov E, Ogievetsky V, Sokatchev E (1986) Hyper-Kähler metrics and harmonic superspace. Commun Math Phys 103:515–526
6. Hitchin N (1992) Hyperkähler manifolds, Sem. Bourbaki, Asterisque 206:137–166
7. Plebanski JF (1975) Some solutions of complex Einstein equations. J Math Phys 16:2395–2402
8. Adamo T, Mason L, Sharma A (2021) Twistor sigma models for quaternionic geometry and graviton scattering. arXiv:2103.16984
9. Pontecorvo M (1992) On twistor spaces of anti-self-dual Hermitian surfaces. Trans Am Math Soc 331:653–661
10. Chakravarty S, Mason L, Newman E (1991) Canonical structures on antiselfdual four manifolds and the diffeomorphism group. J Math Phys 32:1458–1464
11. Salamon S (1982) Quaternionic Kähler manifolds. Invent Math 67:143–171
12. Swann A (1991) Hyper-Kähler and quaternionic Kähler geometry. Math Ann 289(3):421–450
13. Przanowski M (1983) Locally hermite Einstein, selfdual gravitational instantons. Acta Phys Polon B 14:625–627
14. Hoegner M (2012) Quaternion-Kaehler four-manifolds and Przanowski's function. J Math Phys 53:103517. arXiv:1205.3977
15. He T, Lysov V, Mitra P, Strominger A (2015) BMS supertranslations and Weinberg's soft graviton theorem. JHEP 05:151. arXiv:1401.7026

Recalling (4.5?) and (4.101), we observe the sequence of conditions

$$[illegible] \quad (4.118)$$

[illegible]

Theorem 4.2 [illegible]

$$[illegible] \quad (4.119)$$

[illegible]

[illegible]

References

[illegible]

Chapter 5
Graviton Scattering in Flat Space

The main applications of twistor sigma models is to the computation of gravitational scattering amplitudes. In four dimensions, gravitational perturbations are classified by whether their linearized Weyl curvatures are self-dual (SD) or anti-self-dual (ASD). These are the positive and negative helicity gravitons. The tree-level gravitational scattering amplitudes are therefore classified by the number of negative (versus positive) helicity external gravitons [1]. Integrability of the purely SD sector means that amplitudes with all positive or all but one positive external gravitons vanish.

The first non-vanishing configuration as one moves away from self-duality is the *maximal helicity violating* (MHV) amplitude: two negative helicity and arbitrarily many positive helicity external gravitons. Explicit, all-multiplicity formulae for tree-level gravitational MHV amplitudes in flat space have been known for decades [2–4], with the optimal formula (in terms of compactness and explicit permutation symmetry) due to Hodges [5]. While the veracity of these formulae is easily established through unitarity techniques (e.g., Berends-Giele or BCFW recursion), a direct first-principles derivation of Hodges' formula from classical general relativity has proven elusive. In this chapter, we show how twistor sigma models together with a new generating functional leads to such a first-principles derivation.

5.1 MHV Generating Functional

The main tool that facilitates our computation of amplitudes is the perturbation theory associated to a first order chiral formulation of general relativity in Euclidean signature. On a general 4-manifold M, such a chiral formulation is expressed in terms of three ASD 2-forms constructed from the tetrad, $\Sigma^{\alpha\beta} = \boldsymbol{\theta}^{\alpha\dot{\alpha}} \wedge \boldsymbol{\theta}^{\beta}{}_{\dot{\alpha}}$, and an ASD spin connection $\Gamma_{\alpha\beta}$ which is a priori independent of the tetrad [6–11]. The classical action functional is:

A. Sharma, *Twistor Sigma Models*, Springer Theses,
https://doi.org/10.1007/978-3-031-50751-9_5

$$S[\boldsymbol{\theta},\, \Gamma] = \int_M \Sigma^{\alpha\beta} \wedge \left(\mathrm{d}\Gamma_{\alpha\beta} + \boldsymbol{\kappa}^2\, \Gamma_\alpha{}^\gamma \wedge \Gamma_{\beta\gamma}\right), \tag{5.1}$$

where $\boldsymbol{\kappa} = \sqrt{16\pi \mathrm{G}}$ is the gravitational coupling constant. The field equations become

$$\mathrm{d}\Sigma^{\alpha\beta} = 2\,\boldsymbol{\kappa}^2\, \Gamma^{(\alpha}{}_\gamma \wedge \Sigma^{\beta)\gamma}\,, \tag{5.2}$$

$$\boldsymbol{\theta}^{\alpha\dot\alpha} \wedge \left(\mathrm{d}\Gamma_{\alpha\beta} + \boldsymbol{\kappa}^2\, \Gamma_\alpha{}^\gamma \wedge \Gamma_{\beta\gamma}\right) = 0 \tag{5.3}$$

which provide a first order formulation of Einstein's vacuum equations.

The first equation of motion (5.2) is just the first structure equation given in (3.5). Hence, it sets $\boldsymbol{\kappa}^2\, \Gamma_{\alpha\beta}$ to equal the ASD spin connection of g. Recalling the decomposition (3.7), the second equation of motion (5.3) then becomes

$$\begin{aligned} &\boldsymbol{\theta}^{\alpha\dot\alpha} \wedge R_{\alpha\beta} = \boldsymbol{\theta}^{\alpha\dot\alpha} \wedge \left(\Phi_{\alpha\beta\dot\gamma\dot\delta}\, \tilde\Sigma^{\dot\gamma\dot\delta} + \frac{R}{12}\, \Sigma_{\alpha\beta}\right) = 0 \\ &\implies \Phi_{\alpha\beta\dot\gamma\dot\delta} = R = 0\,, \end{aligned} \tag{5.4}$$

where we have used the identity $\boldsymbol{\theta}^{(\alpha|\dot\alpha} \wedge \Sigma^{|\gamma\delta)} = 0$ to drop a term containing the Weyl spinor. This is equivalent to Ricci flatness $R_{ab} = 0$. Thus, we land on the vacuum equations of GR. Similarly, one can also check that off-shell, after integrating out its quadratic dependence on $\Gamma_{\alpha\beta}$, (5.1) equals the Einstein-Hilbert action up to a topological term. Hence, it is equivalent to the Einstein-Hilbert action in perturbation theory.

As $\boldsymbol{\kappa} \to 0$, an integration by parts reduces the chiral action to

$$\lim_{\boldsymbol{\kappa}\to 0} S[\boldsymbol{\theta},\, \Gamma] = \int_M \Gamma_{\alpha\beta} \wedge \mathrm{d}\Sigma^{\alpha\beta}\,. \tag{5.5}$$

In this case, the field $\Gamma_{\alpha\beta}$ stops being a spin connection. Instead, it acts as a Lagrange multiplier imposing $\mathrm{d}\Sigma^{\alpha\beta} = 0$, i.e., flatness of the ASD spin connection. Hence, $\boldsymbol{\kappa} \to 0$ is the limit in which we obtain an action for self-dual GR [12]. At non-zero $\boldsymbol{\kappa}$, (5.1) expands the non-self-dual theory perturbatively around its self-dual subsector. The $O(\boldsymbol{\kappa}^2)$ term in (5.1) encodes graviton interaction vertices and scattering amplitudes [13].

Linearized gravity around SD backgrounds. Graviton perturbations $(\delta\boldsymbol{\theta}^{\alpha\dot\alpha}, \delta\Gamma_{\alpha\beta}) = (\vartheta^{\alpha\dot\alpha}, \gamma_{\alpha\beta})$ around an SD vacuum background M with tetrad $\boldsymbol{\theta}_0^{\alpha\dot\alpha}$ and closed ASD 2-forms $\Sigma_0^{\alpha\beta}$ satisfy the linearized Einstein vacuum equations

$$\mathrm{d}\sigma^{\alpha\beta} = 2\,\boldsymbol{\kappa}^2\, \gamma^{(\alpha}{}_\delta \wedge \Sigma_0^{\beta)\delta}\,, \qquad \boldsymbol{\theta}_0^{\alpha\dot\alpha} \wedge \mathrm{d}\gamma_{\alpha\beta} = 0\,, \tag{5.6}$$

where $\sigma^{\alpha\beta} \equiv \delta\Sigma^{\alpha\beta} = 2\,\boldsymbol{\theta}_0^{(\alpha|\dot\alpha} \wedge \vartheta^{|\beta)}{}_{\dot\alpha}$. The identity $\boldsymbol{\theta}_0^{(\alpha|\dot\alpha} \wedge \Sigma_0^{|\gamma\delta)} = 0$ implies that the second of these field equations is equivalent to linearized Ricci flatness,

$$\delta\Phi_{\alpha\beta\dot\alpha\dot\beta} = \delta R = 0 \implies \mathrm{d}\gamma_{\alpha\beta} = \psi_{\alpha\beta\gamma\delta}\,\Sigma_0^{\gamma\delta}\,, \tag{5.7}$$

where $\psi_{\alpha\beta\gamma\delta}$ is the perturbation to the ASD Weyl spinor.

On SD backgrounds, we can classify such perturbations as having positive or negative helicity. Perturbations preserving self-duality satisfy $\psi_{\alpha\beta\gamma\delta} = 0$ and are referred to as *positive helicity gravitons*. Up to linearized gauge transformations $\delta\gamma_{\alpha\beta} = \mathrm{d}\rho_{\alpha\beta}$, we can take the corresponding solutions of (5.6) and (5.7) to have $\gamma_{\alpha\beta} = 0$ and satisfy $\mathrm{d}\sigma^{\alpha\beta} = 0$. On the other hand, *negative helicity gravitons* are defined by quotienting the space of all solutions of the linearized Einstein equations by the space of such SD solutions [14]. Beyond flat space, they are no longer strictly anti-self-dual and can lead to non-trivial perturbations of $\tilde\Psi_{\dot\alpha\dot\beta\dot\gamma\dot\delta}$. Nevertheless, they will be explicitly computable in the examples we study.

Generating functional for MHV amplitudes. In this thesis, we will illustrate our techniques with a focus on tree-level maximally helicity violating (MHV) amplitudes. These are the amplitudes for scattering two negative helicity gravitons off an arbitrary number of positive helicity gravitons. On SD backgrounds, tree amplitudes with less than two negative helicity gravitons necessarily vanish due to the integrability of SD GR, so the MHV configuration is the first non-trivial case of interest.

Let us consider an n-graviton MHV amplitude $A(1^- 2^- 3^+ \cdots n^+)$ involving negative helicity gravitons (ϑ_i, γ_i), $i = 1, 2$, and positive helicity gravitons $(\vartheta_i, \gamma_i = 0)$, $i = 3, \dots, n$. In [14], it was shown that this can be generated from a two-point function of the negative helicity gravitons on a deformed SD vacuum background $\mathcal{M}$. As a smooth manifold, $\mathcal{M}$ is the same as M. But the metric on $\mathcal{M}$ is constructed as a deformation of the metric on M by a superposition of the positive helicity gravitons. Explicitly, the tetrad and ASD 2-forms on $\mathcal{M}$ take the form

$$\boldsymbol{\theta}^{\alpha\dot\alpha} = \boldsymbol{\theta}_0^{\alpha\dot\alpha} + \sum_{i=3}^{n} a_i\vartheta_i^{\alpha\dot\alpha} + O(a_i a_j \vartheta_i\vartheta_j)\,, \qquad \Sigma^{\alpha\beta} = \boldsymbol{\theta}^{\alpha\dot\alpha}\wedge\boldsymbol{\theta}^{\beta}{}_{\dot\alpha}\,, \tag{5.8}$$

where the higher order corrections are to be determined by solving the fully non-linear SD vacuum equations $\mathrm{d}\Sigma^{\alpha\beta} = 0$ perturbatively in the linear solutions $a_i\vartheta_i$. The a_i are formal parameters that act as classical analogues of 'creation operators' and always accompany the corresponding ϑ_i.

The two-point function of negative helicity gravitons 1, 2 on SD vacuum space-times was determined directly from the action (5.1) in [14][1]:

$$\mathcal{G}(1,2) \equiv \int_{\mathcal{M}} \Sigma^{\alpha\beta}\wedge\gamma_{1\,\alpha}{}^{\delta}\wedge\gamma_{2\,\delta\beta}\,. \tag{5.9}$$

Intuitively, it arises from the interaction term of (5.1) evaluated on-shell in our scattering configuration. So it provides the leading correction about the SD sector.

[1] Their analysis was originally performed in Lorentzian signature. The final generating functional (5.9) can nevertheless be calculated in Euclidean signature and then analytically continued to other signatures.

The n-point MHV amplitude is then given by the term in $\mathcal{G}(1,2)$ that is linear in each a_i,

$$A(1^- 2^- 3^+ \cdots n^+) = \left(\frac{\partial^{n-2}}{\partial a_3 \cdots \partial a_n} \mathcal{G}(1,2) \right) \bigg|_{a_i = 0\ \forall i} . \tag{5.10}$$

This way of computing amplitudes is closely related to perturbiner methods [15]. Intuitively, the $\partial/\partial a_i$ in this expression act as classical analogues of 'annihilation operators'. In particular, picking this multi-linear piece ensures that the resulting amplitude acts as a linear functional on the graviton Hilbert space (after Wick rotation to Lorenztian signature).

5.2 Uplift to Twistor Space

To begin our foray into perturbative gravity, we wish to compute graviton MHV amplitudes on a flat background $M = \mathbb{R}^4$ by treating it as a self-dual background. Sadly, even on $\mathbb{R}^4$, computing (5.8) and (5.9) using space-time perturbation theory quickly becomes untenable due to an exponential proliferation of Feynman diagrams. To overcome this, we will automate the perturbative expansion in (5.8) using our twistor sigma models.

Momentum eigenstates. We will study the scattering of graviton momentum eigenstates labeled by null momenta $k_i^{\alpha\dot\alpha} = \kappa_i^\alpha \tilde\kappa_i^{\dot\alpha}$. The variables $\kappa_i, \tilde\kappa_i$ are known as spinor-helicity variables in the literature. They are defined up to little group scalings

$$(\kappa_i, \tilde\kappa_i) \sim (s_i \kappa_i, s_i^{-1} \tilde\kappa_i)\,, \qquad s_i \in \mathbb{C}^*\,. \tag{5.11}$$

Of course, there are no real null momenta in Euclidean signature. Thankfully, we can easily Wick rotate the resulting amplitudes to Lorentzian signature by setting $\tilde\kappa_i^{\dot\alpha} = \bar\kappa_i^{\dot\alpha}$.

As described in the last section, we will refine our scattering problem by helicity. Let η denote the flat metric (irrespective of signature). On flat space, positive and negative helicity metric perturbations $g = \eta + h^\pm$ associated to momentum eigenstates have the well-known expressions (the factors of -2 are a useful convention)

$$h^+_{i\,\alpha\dot\alpha\beta\dot\beta}(x) = -\frac{2\,\xi_\alpha\,\xi_\beta\,\tilde\kappa_{i\dot\alpha}\,\tilde\kappa_{i\dot\beta}}{\langle \xi\, i\rangle^2}\,\mathrm{e}^{\mathrm{i}k_i\cdot x}\,, \qquad h^-_{i\,\alpha\dot\alpha\beta\dot\beta}(x) = -\frac{2\,\kappa_{i\alpha}\,\kappa_{i\beta}\,\tilde\xi_{\dot\alpha}\,\tilde\xi_{\dot\beta}}{[\tilde\xi\, i]^2}\,\mathrm{e}^{\mathrm{i}k_i\cdot x}\,, \tag{5.12}$$

where ξ_α, $\tilde\xi_{\dot\alpha}$ are arbitrary 'reference spinors' that describe the linearized gauge freedom in the perturbations; and $\langle \xi\, i\rangle \equiv \langle \xi\, \kappa_i\rangle$, $[\tilde\xi\, i] = [\tilde\xi\, \tilde\kappa_i]$, etc. Amplitudes $A(1\,2\cdots n)$ of such momentum and helicity eigenstates transform as

$$A(s_i\kappa_i, s_i^{-1}\tilde\kappa_i) = s_i^{-2J_i}\, A(\kappa_i, \tilde\kappa_i) \tag{5.13}$$

under little group scalings of the ith-graviton with helicity $J_i \in \{\pm 2\}$. This is inherited from little group scalings of the perturbations (5.12), as the amplitudes are linear functionals of these wavefunctions.

On perturbing $\eta \mapsto g = \eta + h$, the perturbations $\vartheta^{\alpha\dot\alpha}$ of the tetrad can be found by linearizing $g = \theta^{\alpha\dot\alpha}\,\theta_{\alpha\dot\alpha}$ around flat space: $\vartheta^{\alpha\dot\alpha} = \frac{1}{2}\,h^{\alpha\dot\alpha}{}_{\beta\dot\beta}\,\mathrm{d}x^{\beta\dot\beta}$. By linearizing (3.4), we can derive the ASD spin connection perturbations generated by the momentum eigenstates of (5.12),

$$\gamma^+_{i\,\alpha\beta} = 0\,, \qquad \gamma^-_{i\,\alpha\beta} = -\mathrm{i}\,\kappa_{i\alpha}\,\kappa_{i\beta}\,\mathrm{e}^{\mathrm{i}k_i\cdot x}\,\frac{\langle i|\mathrm{d}x|\tilde\xi]}{[\tilde\xi\, i]}\,, \tag{5.14}$$

where we have introduced the notation $\langle a|p|b] \equiv a_\alpha p^{\alpha\dot\alpha} b_{\dot\alpha}$ for the contraction of a vector $p^{\alpha\dot\alpha}$ with a pair of spinors a_α, $b_{\dot\alpha}$. Using (5.7), we find the curvature perturbations

$$\psi^+_{i\,\alpha\beta\gamma\delta} = 0\,, \qquad \psi^-_{i\,\alpha\beta\gamma\delta} = \kappa_{i\alpha}\,\kappa_{i\beta}\,\kappa_{i\gamma}\,\kappa_{i\delta}\,\mathrm{e}^{\mathrm{i}k_i\cdot x}\,. \tag{5.15}$$

As expected, the + helicity states are self-dual; and the – helicity states are anti-self-dual by the parity symmetry of flat space.

From generating functional to twistor sigma model. Consider the MHV amplitude $A(1^-\,2^-\,3^+\cdots n^+)$ of momentum eigenstates (5.12). It is computed by the generating functional (5.9). The relevant deformed manifold $\mathcal{M}$ is topologically $\mathbb{R}^4$, but now comes equipped with a deformed metric

$$g = \eta + \sum_{i=3}^{n} a_i h_i^+ + O(a_i a_j h_i^+ h_j^+) \tag{5.16}$$

that solves the self-dual vacuum equations $\mathrm{d}\Sigma^{\alpha\beta} = 0$.

For ease of notation, suppose $\langle 1\,2\rangle \equiv \langle\kappa_1\,\kappa_2\rangle = -1$; factors of $\langle 1\,2\rangle$ will be reinstated in the final amplitudes using little group scaling. Choosing $\kappa_{1\alpha}$, $\kappa_{2\alpha}$ as a basis of undotted spinors, we can define complex coordinates on $\mathcal{M}$,

$$y^{\dot\alpha} = x^{\alpha\dot\alpha}\kappa_{1\alpha}\,, \qquad \tilde y^{\dot\alpha} = x^{\alpha\dot\alpha}\kappa_{2\alpha}\,, \tag{5.17}$$

as we did on a generic SD vacuum space-time in (3.42). Recall from (3.9) that since $\mathcal{M}$ is hyperkähler, we can take its basis of ASD 2-forms to be

$$\Sigma^{11} = \mathrm{d}y^{\dot\alpha}\wedge\mathrm{d}y_{\dot\alpha}\,, \qquad \Sigma^{12} = -\frac{\partial^2\Omega}{\partial y^{\dot\alpha}\partial\tilde y^{\dot\beta}}\,\mathrm{d}y^{\dot\alpha}\wedge\mathrm{d}\tilde y^{\dot\beta}\,, \qquad \Sigma^{22} = \mathrm{d}\tilde y^{\dot\alpha}\wedge\mathrm{d}\tilde y_{\dot\alpha}\,, \tag{5.18}$$

such that the Kähler scalar Ω satisfies Plebanski's heavenly equation (3.13).

In the $y^{\dot\alpha}$, $\tilde y^{\dot\alpha}$ coordinates, the negative helicity spin connection perturbations (5.14) become

$$\gamma_1^{\alpha\beta} = \mathrm{i}\,\kappa_1^\alpha\,\kappa_1^\beta\,\mathrm{e}^{\mathrm{i}[y\,1]}\,\frac{[\tilde{\xi}\,\mathrm{d}y]}{[\tilde{\xi}\,1]}\,, \qquad \gamma_2^{\alpha\beta} = \mathrm{i}\,\kappa_2^\alpha\,\kappa_2^\beta\,\mathrm{e}^{\mathrm{i}[\tilde{y}\,2]}\,\frac{[\tilde{\xi}\,\mathrm{d}\tilde{y}]}{[\tilde{\xi}\,2]}\,. \tag{5.19}$$

The MHV generating functional (5.9) can then be expressed as:

$$\begin{aligned}\int_{\mathcal{M}} \Sigma^{\alpha\beta}\wedge\gamma_{1\,\alpha}{}^{\delta}\wedge\gamma_{2\,\delta\beta} &= \int_{\mathcal{M}} \mathrm{d}^2y\,\mathrm{d}^2\tilde{y}\,\frac{\partial^2\Omega}{\partial y^{\dot\alpha}\partial\tilde{y}^{\dot\beta}}\,\frac{\tilde{\xi}^{\dot\alpha}\,\tilde{\xi}^{\dot\beta}}{[\tilde{\xi}\,1]\,[\tilde{\xi}\,2]}\,\mathrm{e}^{\mathrm{i}[y\,1]+\mathrm{i}[\tilde{y}\,2]}\,,\\ &= -\int_{\mathcal{M}} \mathrm{d}^2y\,\mathrm{d}^2\tilde{y}\;\Omega\,\mathrm{e}^{\mathrm{i}[y\,1]+\mathrm{i}[\tilde{y}\,2]}\,,\end{aligned} \tag{5.20}$$

upon integration by parts in y and $\tilde{y}$. Quite satisfyingly, this generating functional is now manifestly independent of the reference spinor $\tilde{\xi}_{\dot\alpha}$. Reexpressing it in the original coordinates $x^{\alpha\dot\alpha}\in\mathbb{R}^4$ gives

$$\mathcal{G}(1,2) = -\int_{\mathbb{R}^4} \mathrm{d}^4x\;\Omega\,\mathrm{e}^{\mathrm{i}(k_1+k_2)\cdot x}\,. \tag{5.21}$$

Hence, the generating functional of momentum space MHV amplitudes is simply a Fourier transform of the Kähler scalar of $\mathcal{M}$.

It is now straightforward to lift the entire MHV generating functional to twistor space, using Theorem 3.2. Applying (3.62), the result is

$$\mathcal{G}(1,2) = \int_{\mathbb{R}^4} \mathrm{d}^2y\,\mathrm{d}^2\tilde{y}\;\mathrm{e}^{\mathrm{i}[y\,1]+\mathrm{i}[\tilde{y}\,2]}\;S_h[m](y)\big|_{\text{on-shell}}\,, \tag{5.22}$$

where $S_h[m](y)$ is the twistor sigma model of the first kind given by (3.49). So, the last step in lifting $\mathcal{G}(1,2)$ to twistor space is determining a twistor representative h for the deformed space-time $\mathcal{M}$. This can be constructed as a superposition of momentum eigenstate twistor representatives.

Twistor representatives. For a spin-2 momentum eigenstate with null momentum $k_i^{\alpha\dot\alpha} = \kappa_i^\alpha\,\tilde{\kappa}_i^{\dot\alpha}$, the twistor representative $h_i \in H^{0,1}(\mathbb{PT},\mathcal{O}(2))$ in complex coordinates $Z^A = (\mu^{\dot\alpha},\lambda_\alpha)$ is given by

$$h_i(Z) = \int_{\mathbb{C}^*} \frac{\mathrm{d}s}{s^3}\wedge\bar{\delta}^2(\kappa_i - s\,\lambda)\,\mathrm{e}^{\mathrm{i}s[\mu\,i]}\,. \tag{5.23}$$

In this expression, the holomorphic delta function is defined as

$$\bar{\delta}^2(\kappa_i - s\,\lambda) := \frac{1}{(2\pi\mathrm{i})^2}\bigwedge_{\alpha=1,2}(\bar{\partial}_{\mathbb{P}^1} + \mathrm{d}\bar{s}\,\partial_{\bar{s}})\left(\frac{1}{\kappa_{i\alpha} - s\,\lambda_\alpha}\right).$$

Using the flat space incidence relations $X:\mu^{\dot\alpha} = x^{\alpha\dot\alpha}\lambda_\alpha$, $h_i(Z)$ pulls back to

$$h_i(x,\lambda) = \int_{\mathbb{C}^*} \frac{\mathrm{d}s}{s^3} \wedge \bar{\delta}^2(\kappa_i - s\,\lambda)\, \mathrm{e}^{\mathrm{i}k_i\cdot x}\,. \tag{5.24}$$

Its Weyl spinor perturbation $\delta\tilde{\Psi}_{\dot\alpha\dot\beta\dot\gamma\dot\delta} \equiv \tilde{\psi}_{\dot\alpha\dot\beta\dot\gamma\dot\delta}$ on space-time can be found from the Penrose integral formula (2.34) for $\ell = +2$:

$$\begin{aligned}\tilde{\psi}^{+}_{i\,\dot\alpha\dot\beta\dot\gamma\dot\delta}(x) &= \int_X \mathrm{D}\lambda \wedge \mathcal{L}_{\partial_{\dot\alpha}}\mathcal{L}_{\partial_{\dot\beta}}\mathcal{L}_{\partial_{\dot\gamma}}\mathcal{L}_{\partial_{\dot\delta}} h_i\big|_X \\ &= \int_{X\times\mathbb{C}^*} \mathrm{D}\lambda \wedge \mathrm{d}s \wedge \bar{\delta}^2(\kappa_i - s\lambda)\; \partial_{\dot\alpha}\partial_{\dot\beta}\partial_{\dot\gamma}\partial_{\dot\delta}(\mathrm{e}^{\mathrm{i}k_i\cdot x}) = \tilde{\kappa}_{i\dot\alpha}\,\tilde{\kappa}_{i\dot\beta}\,\tilde{\kappa}_{i\dot\gamma}\,\tilde{\kappa}_{i\dot\delta}\,\mathrm{e}^{\mathrm{i}k_i\cdot x}\end{aligned} \tag{5.25}$$

where the integrals were localized against the holomorphic delta functions.

We now take a linear superposition of such representatives,

$$h = \sum_{i=3}^{n} a_i h_i\,. \tag{5.26}$$

Since $\bar\partial h = 0$ (by linearity), it acts as a representative for the linearized graviton $h_{\alpha\dot\alpha\beta\dot\beta} = \sum_{i=3}^{n} a_i h^{+}_{i\,\alpha\dot\alpha\beta\dot\beta}$ on space-time. Moreover, since each $h_i \propto \bar{e}^0$ (due to $\bar\partial_{\mathbb{P}^1} = \bar{e}^0\,\bar\partial_0$) we observe that $\{h, h\} \propto \bar{e}^0 \wedge \bar{e}^0 = 0$. Thus, h also satisfies $T = \bar\partial h + \frac{1}{2}\{h, h\} = 0$ and gives a fully non-linear representative for an integrable complex structure deformation of $\mathbb{PT}$. Applying the non-linear graviton Theorem 3.1 to this representative, we can construct the hyperkähler space-time $\mathcal{M}$ whose metric is asymptotically described by (5.16).

This is the beauty of the non-linear graviton construction. On space-time, it appears like a nigh impossible task to find the higher order corrections in (5.16) that ensure $\mathrm{d}\Sigma^{\alpha\beta} = 0$. But on twistor space, (5.26) solves both the 'linear field equation' $\bar\partial h = 0$ as well as its non-linear counterpart $T = 0$. A non-linear superposition of positive helicity gravitons on space-time reduces to a linear superposition on twistor space!

5.3 MHV Amplitudes from Sigma Model

Sigma model correlators. To extract the amplitude, we need to extract the part of the generating functional which is multi-linear in the positive helicity gravitons. This problem is easily translated into extracting a connected, tree-level correlation function from the field theory on $\mathbb{P}^1$ defined by the twistor sigma model:

Lemma 5.1 *Let $h = \sum_{i=3}^{n} a_i h_i$. When $m^{\dot\alpha}(y, \lambda)$ is a solution to its equation of motion and vanishes at κ_1, κ_2, there is an equivalence*

$$\left(\prod_i \frac{\partial}{\partial a_i}\right) \int_{\mathbb{P}^1} \frac{\mathrm{D}\lambda \wedge \bar{e}^0}{2\pi \mathrm{i}\, \lambda_1^2 \lambda_2^2} \left(\frac{1}{2} \left[m\, \bar{\partial}_0 m \right] + h|_Y \right) \Bigg|_{a_i=0} = \left\langle V_{h_3} V_{h_4} \dots V_{h_n} \right\rangle^0_{\text{tree}} , \tag{5.27}$$

where $\lambda_1 = -\langle \lambda\, 2 \rangle$, $\lambda_2 = \langle \lambda\, 1 \rangle$; $Y : \mu^{\dot\alpha} = \lambda_1 y^{\dot\alpha} + \lambda_2 \tilde{y}^{\dot\alpha} + m^{\dot\alpha}(y, \lambda)$ *are the perturbed holomorphic curves; and the 'vertex operators'* V_{h_i} *are defined as*

$$V_{h_i} = \frac{1}{2\pi \mathrm{i}} \int_{\mathbb{P}^1} \frac{h_i|_Y\, \mathrm{D}\lambda}{\lambda_1^2 \lambda_2^2} , \tag{5.28}$$

and $\langle \dots \rangle^0_{\text{tree}}$ *denotes the correlation function obtained using connected, tree-level Feynman diagrams of the twistor sigma model with trivial background* $h = 0$.

Proof Let $\hbar$ be a formal parameter. The generating functional of connected correlation functions of such vertex operators is the effective action

$$W(a_i) = -2\pi \mathrm{i} \hbar \log \int \mathcal{D}m \, \exp \left(-\frac{1}{4\pi \mathrm{i} \hbar} \int_{\mathbb{P}^1} \frac{\mathrm{D}\lambda \wedge \bar{e}^0}{\lambda_1^2 \lambda_2^2} \left[m\, \bar{\partial}_0 m \right] - \frac{1}{\hbar} \sum_i a_i V_{h_i} \right) , \tag{5.29}$$

as can be inductively confirmed by differentiation with respect to the a_i. Contributions of the connected tree-level graphs can be extracted through the $\hbar \to 0$ limit. Using the saddle point approximation, this reduces to the corresponding on-shell action

$$\begin{aligned} W^{\text{tree}}(a_i) = \lim_{\hbar \to 0} W(a_i) &= \frac{1}{2} \int_{\mathbb{P}^1} \frac{\mathrm{D}\lambda \wedge \bar{e}^0}{\lambda_1^2 \lambda_2^2} \left[m\, \bar{\partial}_0 m \right] + \sum_i a_i V_{h_i} \Bigg|_{\text{on-shell}} \\ &= \int_{\mathbb{P}^1} \frac{\mathrm{D}\lambda \wedge \bar{e}^0}{\lambda_1^2 \lambda_2^2} \left(\frac{1}{2} \left[m\, \bar{\partial}_0 m \right] + h|_Y \right) \Bigg|_{\text{on-shell}} , \end{aligned} \tag{5.30}$$

where now $h = \sum_i a_i h_i$. □

To compute the correlators, we substitute $m^{\dot\alpha} = \lambda_1 \lambda_2\, \mathrm{m}^{\dot\alpha}$ to get the free theory action:

$$\frac{1}{4\pi \mathrm{i}} \int_{\mathbb{P}^1} \frac{\mathrm{D}\lambda \wedge \bar{e}^0}{\lambda_1^2 \lambda_2^2} \left[m\, \bar{\partial}_0 m \right] = \frac{1}{4\pi \mathrm{i}} \int_{\mathbb{P}^1} \mathrm{D}\lambda \wedge \bar{e}^0 \left[\mathrm{m}\, \bar{\partial}_0 \mathrm{m} \right] \tag{5.31}$$

with the only boundary conditions on the $\mathcal{O}(-1)$-valued functions $\mathrm{m}^{\dot\alpha}$ being that they be regular at κ_1, κ_2. This is the action we use to compute OPEs of the vertex operators V_{h_i}. At this point, the computation of this connected tree correlator follows straightforwardly by an application of the weighted matrix-tree theorem.

Lemma 5.2 *The connected tree-level correlators are given by*

$$\left\langle V_{h_3} V_{h_4} \dots V_{h_n} \right\rangle^0_{\text{tree}} = \int_{(\mathbb{P}^1)^{n-2}} \left(|\mathcal{L}^j_j| \prod_{i=3}^{n} \frac{h_i(\mathrm{m}_i)\, \mathrm{D}\lambda_i}{\langle \lambda_i\, 1 \rangle^2 \langle \lambda_i\, 2 \rangle^2} \right) \Bigg|_{\mathrm{m}^{\dot\alpha}=0} , \tag{5.32}$$

where: $\mathrm{m}_i^{\dot\alpha} = \mathrm{m}^{\dot\alpha}(y, \lambda_i)$; V_{h_i} *are as defined in* (5.28)*; the* $(n-2)\times(n-2)$ *matrix* $\mathcal{L}$ *has entries*

$$\mathcal{L}_{ij} = \begin{cases} \frac{1}{\langle\lambda_i\,\lambda_j\rangle}\left[\frac{\partial}{\partial \mathrm{m}_i}\,\frac{\partial}{\partial \mathrm{m}_j}\right], & i \neq j\,, \\ -\sum_{k\neq i}\frac{1}{\langle\lambda_i\,\lambda_k\rangle}\left[\frac{\partial}{\partial \mathrm{m}_i}\,\frac{\partial}{\partial \mathrm{m}_k}\right], & i = j\,, \end{cases} \tag{5.33}$$

and $|\mathcal{L}^j_j|$ *is the determinant of* $\mathcal{H}$ *with one row and column removed, corresponding to any* $j \in \{3, \ldots, n\}$.

Proof Viewing the $\{V_{h_i}\}$ as insertions at n points on $\mathbb{P}^1$, each labeled by homogeneous coordinate λ_i^α, we must extract the sum of connected tree-level Feynman diagrams in the theory defined by the background twistor sigma model with $h = 0$. The relevant kinetic term in the twistor sigma model is $[\mathrm{m}\,\bar\partial_0\mathrm{m}]$, which means that the OPE of a pair of m's inserted at λ_i, λ_j is determined by inverting the $\bar\partial$-operator on sections of $\mathcal{O}(-1)\to\mathbb{P}^1$:

$$\mathrm{m}_i^{\dot\alpha}\,\mathrm{m}_j^{\dot\beta} \sim \frac{2\pi\mathrm{i}\,\epsilon^{\dot\alpha\dot\beta}}{\langle\lambda_i\,\lambda_j\rangle}\,. \tag{5.34}$$

The OPE between any two vertex operator insertions are then determined by acting with $\mathrm{m}^{\dot\alpha}$-derivatives in the appropriate way:

$$h_i(\mathrm{m}_i)\,h_j(\mathrm{m}_j) \sim \frac{2\pi\mathrm{i}}{\langle\lambda_i\,\lambda_j\rangle}\left[\frac{\partial h_i(\mathrm{m}_i)}{\partial \mathrm{m}_i}\,\frac{\partial h_j(\mathrm{m}_j)}{\partial \mathrm{m}_j}\right], \tag{5.35}$$

having dropped terms with multiple Wick contractions as they do not contribute to tree diagrams. The factors of $2\pi\mathrm{i}$ in such OPEs always cancel factors of $1/2\pi\mathrm{i}$ in the vertex operator normalizations at tree-level, so can be freely dropped (Fig. 5.1).

The weighted matrix tree theorem of algebraic combinatorics (cf., [16–18]) states that the sum of all connected tree-level Feyman diagrams is given by the determinant of the weighted Laplacian matrix for the configuration of vertex operators with a row and column corresponding to any one of the vertex operators, say $j \in \{3, \ldots, n\}$ removed. The theorem guarantees that the sum is independent of this choice. From

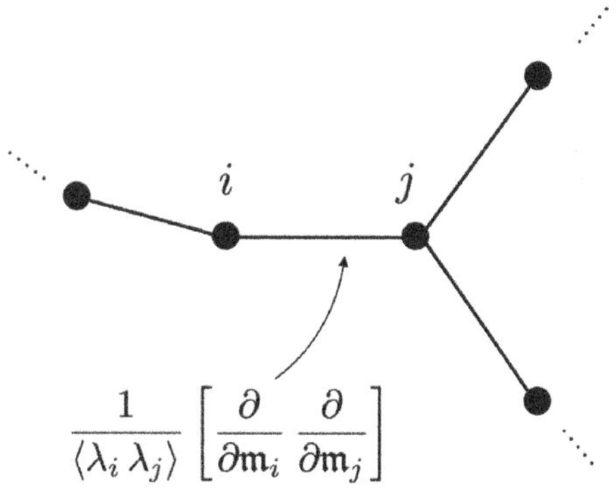

Fig. 5.1 A typical tree diagram contributing to the sigma model correlator. The vertices denote + helicity graviton vertex operators, and edges denote sigma model propagators

(5.34), it follows that the weighted Laplacian matrix is given by $\mathcal{L}$ with entries as in (5.33), meaning that the connected tree correlator takes the claimed form. Lastly, having performed all Wick contractions, we set the perturbations $\mathrm{m}^{\dot\alpha} = 0$ as they have no zero-modes. □

When momentum eigenstates are used, (5.35) yields a factor of $[ij]/\langle ij\rangle$, where $[i\, j] = [\tilde\kappa_i\, \tilde\kappa_j]$, etc. This was the propagator for the tree-diagram formalism of [3, 4]. The vertices similarly provide simple weight factors that can be identified with those of [3, 4]. See [19–21] for analogous usages of the matrix tree theorem to sum these tree diagrams.

Hodges' formula. On inserting momentum eigenstates as wave functions into our general formulae (5.27), our sum of tree diagrams reduces to those of [3, 4]. Its sum via a matrix tree argument (5.32) reduces to Hodges' determinant formula [5] following [19].

Lemma 5.2 combined with (5.22) provides an explicit twistorial formula for the n-point, tree-level graviton MHV amplitude in flat space:

$$\langle 1\, 2\rangle^6 \int\limits_{\mathbb{M}\times(\mathbb{P}^1)^{n-2}} \mathrm{d}^4x\; \mathrm{e}^{\mathrm{i}(k_1+k_2)\cdot x} \left(\left|\mathcal{L}^j_j\right| \prod_{i=3}^{n} h_i\, \mathrm{D}\lambda_i \right)\Bigg|_{\mathrm{m}^{\dot\alpha}=0}, \tag{5.36}$$

where an extra factor of $\langle 1\, 2\rangle^6$ has been reinstated by demanding little group scaling. Each of the twistor wavefunctions is a momentum eigenstate of the form (5.23). Consequently, (5.36) is equal to

$$\langle 1\, 2\rangle^6 \int\limits_{\mathbb{M}\times(\mathbb{P}^1)^{n-2}} \mathrm{d}^4x\; \mathrm{e}^{\mathrm{i}(k_1+k_2)\cdot x} \tag{5.37}$$
$$\times \left\{ \left|\mathcal{L}^j_j\right| \prod_{i=3}^{n} \frac{\mathrm{D}\lambda_i}{\langle\lambda_i\, 1\rangle^2\, \langle\lambda_i\, 2\rangle^2} \int_{\mathbb{C}^*} \frac{\mathrm{d}s_i}{s_i^3}\, \bar\delta^2(\kappa_i - s_i\, \lambda_i)\, \mathrm{e}^{\mathrm{i}s_i\langle\lambda_i|x|i] + \mathrm{i}s_i\langle\lambda_i\, 1\rangle\langle\lambda_i\, 2\rangle[\mathrm{m}_i\, i]} \right\}\Bigg|_{\mathrm{m}^{\dot\alpha}=0},$$

when evaluated on momentum eigenstates. Now, using the fact that

$$\frac{\partial}{\partial \mathrm{m}_i^{\dot\alpha}} = \mathrm{i}\, s_i\, \langle\lambda_i\, 1\rangle\langle\lambda_i\, 2\rangle\, \tilde\kappa_{i\dot\alpha}\,, \tag{5.38}$$

when acting on $\mathrm{e}^{\mathrm{i}s_i\langle\lambda_i\, 1\rangle\langle\lambda_i\, 2\rangle[\mathrm{m}_i\, i]}$, it follows that

$$\left|\mathcal{L}^j_j\right| \prod_{k\neq 1,2,j} \langle\lambda_k\, 1\rangle^{-2}\, \langle\lambda_k\, 2\rangle^{-2} = \left|\mathcal{H}^j_j\right|, \tag{5.39}$$

where $\mathcal{H}$ is the $(n-2)\times(n-2)$ matrix whose entries are

$$\mathcal{H}_{ij} = \begin{cases} s_i s_j \, \frac{[i\,j]}{\langle \lambda_i\, \lambda_j \rangle}\,, & i \neq j\,, \\ -s_i \sum_{k\neq i} s_k \frac{[i\,k]}{\langle \lambda_i\, \lambda_k \rangle} \frac{\langle \lambda_k\, 1\rangle \langle \lambda_k\, 2\rangle}{\langle \lambda_i\, 1\rangle \langle \lambda_i\, 2\rangle}\,, & i = j\,, \end{cases} \tag{5.40}$$

and $|\mathcal{H}^j_j|$ is the determinant with the row and column corresponding to graviton j removed.

The $n-2$ integrations over $\lambda_i^\alpha \in \mathbb{P}^1$ can be performed against holomorphic delta functions in the momentum eigenstate representatives, which set $s_i\,\lambda_i = \kappa_i$. The remaining scale integrals over the s_i parameters are also trivially performed against the holomorphic delta functions in the representatives. The final result is:

$$\frac{\langle 1\,2\rangle^6}{\langle 1\,j\rangle^2\,\langle 2\,j\rangle^2}\,|\mathbb{H}^j_j| \int_{\mathbb{R}^4} \mathrm{d}^4x\, \exp\left[\mathrm{i}\sum_{i=1}^n k_i\cdot x\right] = \delta^4\!\left(\sum_{i=1}^n k_i\right) \frac{\langle 1\,2\rangle^6}{\langle 1\,j\rangle^2\,\langle 2\,j\rangle^2}\,|\mathbb{H}^j_j|\,, \tag{5.41}$$

where the $(n-2)\times(n-2)$ matrix $\mathbb{H}$—called the Hodges matrix—has entries

$$\begin{aligned} \mathbb{H}_{ij} &= \frac{[i\,j]}{\langle i\,j\rangle}\,, \qquad i\neq j \in \{3,\ldots,n\}\,, \\ \mathbb{H}_{ii} &= -\sum_{j\neq i} \frac{[i\,j]}{\langle i\,j\rangle}\,\frac{\langle 1\,j\rangle\,\langle 2\,j\rangle}{\langle 1\,i\rangle\,\langle 2\,i\rangle}\,, \qquad i\in\{3,\cdots,n\}\,, \end{aligned} \tag{5.42}$$

and $|\mathbb{H}^j_j|$ is the determinant of the matrix obtained by removing the jth row and jth column from it. This is precisely Hodges' formula for the n-point graviton MHV amplitude [5], where gravitons 1 and 2 are negative helicity and all others are positive helicity. The fact that the combination

$$\frac{|\mathbb{H}^j_j|}{\langle 1\,j\rangle^2\,\langle 2\,j\rangle^2} \tag{5.43}$$

is independent of the choice of positive helicity graviton j follows from the identity

$$\sum_{j=3}^n \mathbb{H}_{ij}\,\langle 1\,j\rangle\,\langle 2\,j\rangle = 0 \qquad \forall\, i \tag{5.44}$$

and properties of determinants.

This provides a first-principles derivation of Hodges' formula by systematically lifting space-time perturbation theory to twistor space. Calculating the Hodges determinant reduces the computational complexity of MHV graviton amplitudes from factorial time (afforded by Feynman diagrams or double copy techniques) to polynomial time. Such remarkable simplifications typically signify the presence of deep symmetries in our description of physics. True to such promise, in the next chapter we will uncover an infinite dimensional symmetry algebra that hides within the dynamics of our twistor sigma models. This will require us to put on the lens of flat space holography.

References

1. Elvang H, Huang Y-T (2013) Scattering amplitudes. arXiv:1308.1697
2. Berends FA, Giele WT, Kuijf H (1988) On relations between multi-gluon and multi-graviton scattering. Phys Lett B 211:91–94
3. Bern Z, Dixon LJ, Perelstein M, Rozowsky JS (1999) Multileg one loop gravity amplitudes from gauge theory. Nucl Phys B 546:423–479 [hep-th/9811140]
4. Nguyen D, Spradlin M, Volovich A, Wen C (2010) The tree formula for MHV graviton amplitudes. JHEP 07:045. arXiv:0907.2276
5. Hodges A (2012) A simple formula for gravitational MHV amplitudes. arXiv:1204.1930
6. Plebanski JF (1977) On the separation of Einsteinian substructures. J Math Phys 18:2511–2520
7. Jacobson T, Smolin L (1987) The left-handed spin connection as a variable for canonical gravity. Phys Lett B 196:39–42
8. Jacobson T, Smolin L (1988) Covariant action for Ashtekar's form of canonical gravity. Class Quant Grav 5:583
9. Mason LJ, Frauendiener J (1990) The sparling 3-form, Ashtekar variables and quasi-local mass, vol 156. LMS Lecture Notes, CUP
10. Capovilla R, Jacobson T, Dell J, Mason LJ (1991) Selfdual two forms and gravity. Class Quant Grav 8:41–57
11. Krasnov K (2011) Plebanski formulation of general relativity: a practical introduction. Gen Rel Grav 43:1–15. arXiv:0904.0423
12. Smolin L (1992) The G(Newton) –> 0 limit of Euclidean quantum gravity. Class Quant Grav 9:883–894. [hep-th/9202076]
13. Abou-Zeid M, Hull CM (2006) A Chiral perturbation expansion for gravity. JHEP 02:057 [hep-th/0511189]
14. Mason LJ, Skinner D (2010) Gravity, twistors and the MHV formalism. Commun Math Phys 294:827–862. arXiv:0808.3907
15. Rosly AA, Selivanov KG (1997) Gravitational SD perturbiner. [hep-th/9710196]
16. Stanley RP (1999) Enumerative combinatorics, vol 2. Cambridge University Press
17. van Lint JH, Wilson RP (2001) A course in combinatorics, vol 2. Cambridge University Press
18. Stanley RP (2012) Topics in algebraic combinatorics. http://www-math.mit.edu/~rstan/algcomb.pdf
19. Feng B, He S (2012) Graphs, determinants and gravity amplitudes. JHEP 10:121. arXiv:1207.3220
20. Adamo T, Mason L (2013) Twistor-strings and gravity tree amplitudes. Class Quant Grav 30:075020. arXiv:1207.3602
21. Adamo T, Mason L (2014) Conformal and Einstein gravity from twistor actions. Class Quant Grav 31:4, 045014. arXiv:1307.5043

Chapter 6
$w_{1+\infty}$ and Celestial Holography

In recent years, there has been a resurgence in the search for holographic dualities in asymptotically flat space-times (cf., [1–4] for recent reviews). This program goes by the name of "celestial holography". Much of the recent work in this direction has focused on the interplay between asymptotic symmetries and soft particles [5, 6]. For example, at leading order in the soft momentum, soft gravitons are related to BMS supertranslations via a Ward identity [7], with many generalizations to subleading orders and other theories (cf., [1] and references therein). In the past, these have been understood to arise from an interplay between asymptotic symmetries and twistor/ambitwistor strings [8–10].

In an exciting turn of events, recent studies of soft and collinear limits of flat space amplitudes [11, 12] have revealed an *infinite* tower of such soft graviton theorems arising by working to all orders in the soft expansion [13–16]. For a positive helicity soft graviton, this infinite tower of soft theorems can be organized into the algebra $w_{1+\infty}$ (or more precisely, the loop algebra of the wedge algebra of $w_{1+\infty}$) [17–20]. It is generated by chiral currents $w[p,q](z)$, with $z \in \mathbb{CP}^1$ and $p, q \in \mathbb{Z}_{\geq 0}$, obeying the operator product expansions

$$w[p,q](z)\, w[r,s](z') \sim \frac{ps-qr}{z-z'}\, w[p+r-1,q+s-1](z')\,. \tag{6.1}$$

It has long been known that the algebra $w_{1+\infty}$ classically describes canonical transformations of a plane [21, 22]. Over the years, a number of authors have linked the corresponding loop algebra (6.1) to self-dual gravity via the non-linear graviton construction [23–26] of deformed twistor spaces for self-dual space-times. As a result, it finds natural links with flat space holography through the work of Newman and Penrose [27–29].

In this chapter, we begin by reviewing the program of celestial holography and the appearance of $w_{1+\infty}$ symmetries. This is followed by a review of how the same symmetries also govern self-dual GR. Finally, we explain how the celestial realization of $w_{1+\infty}$ arises as a consequence of the SD GR realization by means of our twistor sigma models.

A. Sharma, *Twistor Sigma Models*, Springer Theses,
https://doi.org/10.1007/978-3-031-50751-9_6

6.1 Symmetries of Celestial CFT

Celestial holography proposes the existence of a 2-dimensional conformal field theory living on a Riemann sphere that is holographically dual to quantum gravity in asymptotically flat space-times. In Lorentzian signature, it can be thought of as living on the celestial sphere at null infinity $\mathscr{I}^+$. Hence, it is dubbed a *celestial conformal field theory*, or CCFT for short.

According to its conjectural holographic dictionary, a graviton momentum eigenstate in 4d flat space is dual to a local operator in CCFT [1, 30, 31]:

$$|\kappa, \tilde{\kappa}, \ell\rangle \quad \longleftrightarrow \quad \mathcal{O}_\ell(\omega, z, \bar{z})\,. \tag{6.2}$$

The subscript $\ell = \pm 2$ denotes graviton helicity, and $\kappa_\alpha = \sqrt{\omega}\,(1, z)$, $\tilde{\kappa}_{\dot\alpha} = \sqrt{\omega}\,(1, \bar{z})$ are spinor-helicity variables (defined up to little group scaling as before). In Lorentzian signature, $\bar{z}$ is the complex conjugate of z, and $z \in \mathbb{C}$ can be identified as a complex coordinate on the celestial sphere. More generally, it is useful to complexify space-time (or work in Euclidean signature as in the previous chapters). In this case, we can think of z and $\bar{z}$ as independent complex variables,[1] and ω as a complex scaling that intuitively encodes the graviton's "energy". Such a complexification is helpful in studying the analytic structure of scattering amplitudes.

Celestial holography posits that graviton amplitudes can be reconstructed as CCFT correlators of the dual operators,

$$A(\omega_i, z_i, \bar{z}_i) = \left\langle \prod_{i=1}^{n} \mathcal{O}_{\ell_i}(\omega_i, z_i, \bar{z}_i) \right\rangle. \tag{6.3}$$

This is particularly natural since Lorentz transformations in the bulk act as 2d conformal transformations on the celestial coordinates $z_i, \bar{z}_i$ and as scalings on ω_i:

$$z_i' = \frac{a z_i + b}{c z_i + d}\,, \qquad \bar{z}_i' = \frac{\bar{a}\bar{z}_i + \bar{b}}{\bar{c}\bar{z}_i + \bar{d}}\,, \qquad \omega_i' = \omega_i\,(c z_i + d)(\bar{c}\bar{z}_i + \bar{d}) \tag{6.4}$$

for $i = 1, \ldots, n$ and $ad - bc = \bar{a}\bar{d} - \bar{b}\bar{c} = 1$. The amplitude is Lorentz covariant and transforms as

$$A(\omega_i', z_i', \bar{z}_i') = \prod_{j=1}^{n} (c z_j + d)^{\ell_j}\,(\bar{c}\bar{z}_j + \bar{d})^{-\ell_j}\,A(\omega_i, z_i, \bar{z}_i)\,. \tag{6.5}$$

[1] CCFT then lives on an abstract $\mathbb{CP}^1$ with coordinate z.

This resembles the transformation law of a conformal correlator, up to the non-trivial transformations of the ω_i's.[2]

As in any CFT, the dynamics of celestial CFT are expected to be governed by its operator product expansions. The OPE limits $z_{ij} \equiv z_i - z_j \to 0$ or $\bar{z}_{ij} \equiv \bar{z}_i - \bar{z}_j \to 0$ when two operators $\mathcal{O}_{\ell_i}(\omega_i, z_i, \bar{z}_i)$, $\mathcal{O}_{\ell_j}(\omega_j, z_j, \bar{z}_j)$ approach each other get reflected in the collinear limits of the corresponding amplitudes [32, 33]. In the limit when two positive helicity gravitons become collinear, tree amplitudes of general relativity factorize as [34]

$$\begin{aligned} &A(\omega_1, z_1, \bar{z}_1, +;\, \omega_2, z_2, \bar{z}_2, +;\, \cdots) \\ &\qquad \sim \frac{(\omega_1+\omega_2)^2}{\omega_1\,\omega_2}\,\frac{\bar{z}_{12}}{z_{12}}\, A\biggl(\omega_1+\omega_2, z_2, \frac{\omega_1\bar{z}_1+\omega_2\bar{z}_2}{\omega_1+\omega_2}, +;\, \cdots\biggr) + O(z_{12}^0) \end{aligned} \tag{6.6}$$

(neglecting factors of the gravitational coupling for brevity's sake). We can make this resemble an OPE expansion by Taylor expanding the amplitude on the right in small $\bar{z}_{12}$,

$$\begin{aligned} &A(\omega_1, z_1, \bar{z}_1, +;\, \omega_2, z_2, \bar{z}_2, +;\, \cdots) \\ &\quad \sim \frac{(\omega_1+\omega_2)^2}{\omega_1\,\omega_2}\,\frac{\bar{z}_{12}}{z_{12}} \sum_{k=0}^{\infty} \frac{1}{k!}\,\frac{\omega_1^k}{(\omega_1+\omega_2)^k}\,\bar{z}_{12}^k\,\partial_{\bar{z}_2}^k A(\omega_1+\omega_2, z_2, \bar{z}_2, +;\, \cdots) + O(z_{12}^0)\,. \end{aligned} \tag{6.7}$$

Thus, we expect the so-called "celestial OPE" of two positive helicity graviton operators to be given by (abbreviating $\mathcal{O}_{+2}$ by $\mathcal{O}_+$)

$$\mathcal{O}_+(\omega_1, z_1, \bar{z}_1)\,\mathcal{O}_+(\omega_2, z_2, \bar{z}_2) \sim \frac{\bar{z}_{12}}{z_{12}} \sum_{k=0}^{\infty} \frac{1}{k!}\,\frac{\omega_1^{k-1}}{\omega_2\,(\omega_1+\omega_2)^{k-2}}\,\bar{z}_{12}^k\,\bar{L}_{-1}^k \mathcal{O}_+(\omega_1+\omega_2, z_2, \bar{z}_2) \tag{6.8}$$

at the leading singular order in z_{12}, where $\bar{L}_{-1} \equiv \partial_{\bar{z}}$ is a 2d conformal generator.

Soft gravitons in CCFT. A graviton with non-zero energy $|\omega| > 0$ is colloquially known as a hard graviton. Similarly, one can define soft gravitons by taking residues of hard gravitons at various orders in the soft limit $\omega \to 0$ [17],

$$w_p(z, \bar{z}) := \operatorname*{Res}_{\omega=0}\, \omega^{1-p}\, \mathcal{O}_\ell(\omega, z, \bar{z})\,, \qquad p \in \mathbb{Z}_{\geq 0}\,. \tag{6.9}$$

Taking the first graviton in (6.8) soft of order p shows that the OPE between $w_p(z_1, \bar{z}_1)$ and an arbitrary hard graviton $O_+(\omega_2, z_2, \bar{z}_2)$ is a polynomial of degree p in $\bar{z}_1$. Hence, assuming that the hard gravitons generate the spectrum of CCFT as

---

[2] On fixing signatures, it is also possible to work with an alternate basis of states where one trades the energy ω for a scaling dimension Δ that is invariant under conformal transformations [12]. This is called a conformal basis. To keep our treatment general, we continue to use the momentum eigenbasis in this work.

an algebra, we can take $w_p(z, \bar{z})$ to be a finite polynomial in $\bar{z}$ (the notation below follows [35]),

$$w_p(z, \bar{z}) = \sum_{k+l=p} \frac{\bar{z}^l \, w[k, l](z)}{k!\,l!} \,. \tag{6.10}$$

This gives rise to $p + 1$ chiral currents $w[k, l](z)$. Taking both gravitons soft in (6.8), one can derive the OPE between these:

$$w[k, l](z_1)\, w[m, n](z_2) \sim \frac{kn - lm}{z_{12}}\, w[k + m - 1, l + n - 1](z_2) \,. \tag{6.11}$$

In [18], Strominger recognized this as the loop algebra of the algebra $w_{1+\infty}$.

To be explicit, assume that $w[k, l](z)$ is meromorphic at $z = 0$ and perform a Laurent expansion

$$w[k, l](z) = \sum_{r \in \mathbb{Z}} \frac{w[k, l]_r}{z^{r+1}} \,. \tag{6.12}$$

From the OPE (6.11), it follows that the modes $w[k, l]_r$ of this current obey the commutation relations

$$\big[w[k, l]_r, w[m, n]_s\big] = (kn - lm)\, w[k + m - 1, l + n - 1]_{r+s} \,, \tag{6.13}$$

where the commutator is to be understood in the sense of a 2d CFT. The zero-modes $w[k, l]_0$ form the algebra $w_{1+\infty}$, the algebra of canonical transformations of a 2-plane (to be discussed shortly). The extra parameter $z \in \mathbb{P}^1$ forms a loop parameter (when restricted to the equator $|z| = 1$), so the algebra (6.13) is known as the loop algebra $Lw_{1+\infty}$. The "1" in $1 + \infty$ stands for the single current $w[0, 0]$ which is a central element of the algebra.

Since gravity is a key part of holographic dualities, we may expect similar symmetry algebras to constrain any celestial CFT. They are commonly referred to as *soft graviton symmetries*. To really understand why such a hidden symmetry algebra exists in the collinear behavior of graviton scattering, we once again turn to self-dual GR.

6.2 Symmetries of Self-dual GR

Canonical transformations and $Lw_{1+\infty}$. Let $\mu^{\dot\alpha} = (\mu^{\dot 1}, \mu^{\dot 2})$ be complex coordinates on $\mathbb{C}^2$, and let $\{\,\cdot\,,\,\cdot\,\}$ be the familiar holomorphic Poisson structure

$$\{f, g\} = \frac{\partial f}{\partial \mu^{\dot 1}} \frac{\partial g}{\partial \mu^{\dot 2}} - \frac{\partial f}{\partial \mu^{\dot 2}} \frac{\partial g}{\partial \mu^{\dot 1}} \,. \tag{6.14}$$

Elements of the Lie algebra $w_{1+\infty}$ of holomorphic canonical transformations can be decomposed into polynomial Hamiltonians on the $\mu^{\dot{\alpha}}$-plane,

$$H[k,l] := (\mu^{\dot{1}})^k\,(\mu^{\dot{2}})^l\,, \qquad k,l \in \mathbb{Z}_{\geq 0}\,. \tag{6.15}$$

The Poisson bracket acting on these elements gives

$$\{H[k,l], H[m,n]\} = \big(kn - lm\big)\,H[k+m-1, l+n-1]\,. \tag{6.16}$$

This defines the commutation relations of the basis elements $H[k,l]$ of $w_{1+\infty}$. Strictly speaking, since $k, l \geq 0$, this is the wedge subalgebra of $w_{1+\infty}$.

The loop algebra $Lw_{1+\infty}$ of $w_{1+\infty}$ can be represented by introducing a complex coordinate $\lambda \in \mathbb{C}$, where the loop is parametrized by $|\lambda| = 1$. Alternatively, λ can be viewed as an affine coordinate on the Riemann sphere $S^2 \simeq \mathbb{P}^1$: if $\lambda_\alpha = (\lambda_1, \lambda_2)$ are homogeneous coordinates on $\mathbb{P}^1$, then on the patch where $\lambda_1 \neq 0$ we can identify $\lambda \equiv \lambda_2/\lambda_1$. With this, the generators of $Lw_{1+\infty}$ can be written as

$$H[k,l]_r := \frac{H[k,l]}{\lambda_1^{k+l+r-2}\,\lambda_2^{-r}} = \lambda^r\,H[k,l]\,, \tag{6.17}$$

where in the second equality we have chosen a scaling for the homogeneous coordinates in which $\lambda_\alpha = (1, \lambda)$. The Poisson bracket (6.14) gives

$$\{H[k,l]_r, H[m,n]_s\} = \big(kn - lm)\,H[k+m-1, l+n-1]_{r+s}\,, \tag{6.18}$$

which define the Lie bracket of the loop algebra $Lw_{1+\infty}$.

Realization on twistor space. In Sect. 3.2, we described the non-linear graviton construction of deformed twistor spaces for self-dual, Ricci-flat 4-manifolds. It provided a correspondence between SD vacuum metrics and hamiltonians $h \in \Omega^{0,1}(\mathbb{PT}, \mathcal{O}(2))$ satisfying the integrability condition $T \equiv \bar{\partial}h + \frac{1}{2}\{h,h\} = 0$.

It is easiest to study the symmetries of this construction in holomorphic coordinates on the deformed twistor space $\mathbb{P}\mathscr{T}$ generated by h. Due to the presence of a holomorphic fibration $\mathbb{P}\mathscr{T} \to \mathbb{P}^1$, the base coordinates $\lambda_\alpha = (\lambda_1, \lambda_2)$ continue to be holomorphic. Cover $\mathbb{P}\mathscr{T} = U_1 \cup U_2$ by open sets

$$U_1 = \{\lambda_1 \neq 0\}\,, \qquad U_2 = \{\lambda_2 \neq 0\}\,. \tag{6.19}$$

Let $\omega^{\dot{\alpha}} \equiv \omega^{\dot{\alpha}}(Z)$ and $\tilde{\omega}^{\dot{\alpha}} \equiv \tilde{\omega}^{\dot{\alpha}}(Z)$ denote holomorphic coordinates on the fibers of $\mathbb{P}\mathscr{T} \to \mathbb{P}^1$ contained in U_1 and U_2 respectively. They are determined by solving

$$\begin{aligned} U_1 \ :& \quad \bar{\partial}\omega^{\dot{\alpha}} + \{h, \omega^{\dot{\alpha}}\} = 0\,, \\ U_2 \ :& \quad \bar{\partial}\tilde{\omega}^{\dot{\alpha}} + \{h, \tilde{\omega}^{\dot{\alpha}}\} = 0\,, \end{aligned} \tag{6.20}$$

subject to the requirements that $\omega^{\dot\alpha}$ be holomorphic in λ_2/λ_1 and $\tilde\omega^{\dot\alpha}$ in λ_1/λ_2. Alternatively, the data contained in h can be repackaged into holomorphic patching functions relating $\omega^{\dot\alpha}$ with $\tilde\omega^{\dot\alpha}$. This was the original approach taken by Penrose in [28].

Any patching functions between $\omega^{\dot\alpha}$ and $\tilde\omega^{\dot\alpha}$ also need to be constrained by further given data on $\mathbb{P}\mathscr{T}$. Recall from the discussion in Sect. 3.2 that the data of a (2, 0)-form $\Sigma = \theta^{\dot\alpha} \wedge \theta_{\dot\alpha}$ satisfying $\mathrm{d}\Sigma = 0 \bmod \mathrm{D}\lambda$ on $\mathbb{P}\mathscr{T}$ encodes the data of SD vacuum metrics on space-time. By Darboux's theorem (by refining our cover if needed), we can choose $\omega^{\dot\alpha}$ and $\tilde\omega^{\dot\alpha}$ such that Σ is in standard form on both patches,

$$\begin{aligned} U_1 &: \quad \Sigma = \mathrm{d}\omega^{\dot\alpha} \wedge \mathrm{d}\omega_{\dot\alpha}\,, \\ U_2 &: \quad \Sigma = \mathrm{d}\tilde\omega^{\dot\alpha} \wedge \mathrm{d}\tilde\omega_{\dot\alpha}\,. \end{aligned} \tag{6.21}$$

On the overlap $U_1 \cap U_2$, equality of the two expressions demands that the patching function be a canonical transformation on each fiber. Infinitesimally, such a canonical transformation is given by

$$\tilde\omega^{\dot\alpha} = \omega^{\dot\alpha} + \frac{\partial \chi}{\partial \omega_{\dot\alpha}} + O(\chi^2)\,, \tag{6.22}$$

where $\chi(Z) \in \Omega^0(U_1 \cap U_2, \mathscr{O}(2))$ is a Hamiltonian that is holomorphic in $\omega^{\dot\alpha}$, λ_α on the overlap.

Holomorphicity on $U_1 \cap U_2$ implies that χ can be Taylor expanded in $\omega^{\dot\alpha}$ and Laurent expanded in λ_α:

$$\chi(\omega, \lambda) = \sum_{k,l \geq 0} \sum_{r \in \mathbb{Z}} c_{k,l,r}\, \frac{(\omega^{\dot 1})^k\, (\omega^{\dot 2})^l}{\lambda_1^{k+l+r-2}\, \lambda_2^{-r}}\,. \tag{6.23}$$

Here, $\lambda_1^{2-k-l-r}\,\lambda_2^r\,(\omega^{\dot 1})^k\,(\omega^{\dot 2})^l$ are just the generators (6.17) of $Lw_{1+\infty}$ with $\mu^{\dot\alpha}$ replaced by the holomorphic coordinates $\omega^{\dot\alpha}$. Thus, the patching function χ is an element of the Lie algebra $Lw_{1+\infty}$ of canonical transformations parametrized by λ_α and preserving the symplectic form $\mathrm{d}\omega^{\dot\alpha} \wedge \mathrm{d}\omega_{\dot\alpha}$ on each fiber. This description of SD GR admits a natural adjoint action of $Lw_{1+\infty}$:

$$\delta\chi = \{\chi, H\}\,, \qquad H \in Lw_{1+\infty}\,. \tag{6.24}$$

On space-time, this action maps SD vacuum metrics to new SD vacuum metrics and descends to the famous infinite dimensional $Lw_{1+\infty}$ symmetry of self-dual GR.

In linear theory around $\mathbb{PT}$, it is standard to view χ as an element of the Čech cohomology group $\check{H}^1(\mathbb{PT}, \mathscr{O}(2))$ [36]. On $U_1 \cap U_2$, any such χ can be expanded on the basis of $Lw_{1+\infty}$ generators $H[k, l]_r$ given in (6.17) in the background holomorphic coordinates $\mu^{\dot\alpha}$. Using the Čech-Dolbeault isomorphism, we can associate every $\chi \in \check{H}^1(\mathbb{PT}, \mathscr{O}(2))$ to a linear perturbation $h \in H^{0,1}(\mathbb{PT}, \mathscr{O}(2))$ of the complex structure. In general, this is done by choosing a partition of unity

$\rho_1 + \rho_2 = 1$ subordinate to the cover $\{U_1, U_2\}$ and defining a pair of smooth functions $\chi_2 = \rho_1\,\chi \in \Omega^0(U_2, \mathcal{O}(2))$ and $\chi_1 = -\rho_2\,\chi \in \Omega^0(U_1, \mathcal{O}(2))$. Assuming that ρ_1, ρ_2 are supported within U_1, U_2 respectively, these are well-defined even though χ is only defined on $U_1 \cap U_2$. Since $\chi = \chi_2 - \chi_1$ is holomorphic, we observe that $\bar\partial\chi_1 = \bar\partial\chi_2$ on the overlap. Hence, the $\bar\partial\chi_i$ patch together to give a smooth element $h \in H^{0,1}(\mathbb{PT}, \mathcal{O}(2))$,

$$h = \rho_1\,\bar\partial\chi_1 + \rho_2\,\bar\partial\chi_2\,. \tag{6.25}$$

(Around a non-trivial background h, one would replace $\bar\partial\chi_i$ by $\bar\partial\chi_i + \{\mathsf{h}, \chi_i\}$.) Note that this is not necessarily globally exact. In this way, we can express elements of $Lw_{1+\infty}$ in our Dolbeault framework. This allows us to make direct contact with soft graviton symmetries.

Soft gravitons in SD GR. In Sect. 6.1, we saw that soft graviton currents $w_p(z, \bar z)$ act as generating functions for the basis $w[k, l]_r$ of the celestial realization of $Lw_{1+\infty}$. Here, we repeat a similar analysis for soft graviton representatives on twistor space.

We start with the momentum eigenstate twistor representative

$$\begin{aligned} h(Z) &= \int_{\mathbb{C}^*} \frac{\mathrm{d}s}{s^3}\,\bar\delta^2(\sqrt{\omega}\,\zeta - s\,\lambda)\,\mathrm{e}^{\mathrm{i}s\sqrt{\omega}[\mu\,\tilde\zeta]} \\ &= \sum_{p=0}^{\infty} \frac{\mathrm{i}^p}{p!}\,\omega^{p-2}\,[\mu\,\tilde\zeta]^p\,\bar\delta_{2-p}(\langle\lambda\,\zeta\rangle)\,, \end{aligned} \tag{6.26}$$

where $k_{\alpha\dot\alpha} = \omega\,\zeta_\alpha\,\tilde\zeta_{\dot\alpha}$ with $\zeta_\alpha = (1, z)$, $\tilde\zeta_{\dot\alpha} = (1, \bar z)$, and we have introduced the notation

$$\bar\delta_m(\langle\lambda\,\kappa\rangle) := \int \frac{\mathrm{d}s}{s^{m+1}}\,\bar\delta^2(\kappa - s\,\lambda) = \left(\frac{\lambda_1}{\kappa_1}\right)^{m+1} \bar\delta(\langle\lambda\,\kappa\rangle) \tag{6.27}$$

for a delta function of weight m in λ_α and $-m-2$ in κ_α.

An order p soft graviton is defined to be the coefficient of ω^{p-2} in this Taylor expansion. This is picked by the same residue prescription (6.9) that we used for celestial operators,

$$h_p(Z) = \operatorname*{Res}_{\omega=0}\,\omega^{1-p}\,h(Z) = \frac{\mathrm{i}^p}{p!}\,[\mu\,\tilde\zeta]^p\,\bar\delta_{2-p}(\langle\lambda\,\zeta\rangle)\,. \tag{6.28}$$

Substituting $[\mu\,\tilde\zeta] = \mu^{\dot 1} + \bar z\,\mu^{\dot 2}$, we can expand this as a polynomial in $\bar z$,

$$h_p(Z) = \sum_{k+l=p} \frac{\bar z^l\,h[k,l](Z)}{k!\,l!}\,, \tag{6.29}$$

$$h[k,l](Z) = \mathrm{i}^{k+l}\,H[k,l]\,\bar\delta_{2-k-l}(\langle\lambda\,\zeta\rangle)\,, \tag{6.30}$$

where $H[k,l] = (\mu^{\dot 1})^k\,(\mu^{\dot 2})^l$ are the $w_{1+\infty}$ Hamiltonians introduced before.

Working in the patch $\lambda_1 \neq 0$, we can use $\bar{\partial}|_{\mathbb{P}^1} \langle\lambda\,\zeta\rangle^{-1} = 2\pi\mathrm{i}\,\bar{\delta}(\langle\lambda\,\zeta\rangle)$ to express (6.30) as

$$h[k,l](Z) = \frac{1}{2\pi\mathrm{i}}\,\bar{\partial}\left(\frac{\mathrm{i}^{k+l}}{\lambda_1^{k+l-3}}\,\frac{H[k,l]}{\langle\lambda\,\zeta\rangle}\right) \tag{6.31}$$

This expresses the soft graviton $h[k,l] = \bar{\partial}\chi[k,l]$ as an $Lw_{1+\infty}$ transformation generated by the Hamiltonian

$$\chi[k,l](Z) = \frac{1}{2\pi\mathrm{i}}\,\frac{\mathrm{i}^{k+l}}{\lambda_1^{k+l-3}}\,\frac{H[k,l]}{\lambda_2 - z\lambda_1}\,, \tag{6.32}$$

having used $\langle\lambda\,\zeta\rangle = \lambda_2 - z\lambda_1$. It is useful to reiterate that $h[k,l]$ is not globally exact, as this Hamiltonian is singular at $\langle\lambda\,\zeta\rangle = 0$. Strictly speaking, we are working with the sheaf of distributional (0, 1)-forms. Next, we can expand $\chi[k,l]$ in z,

$$\chi[k,l](Z) = \begin{cases} (2\pi\mathrm{i})^{-1}\sum_{r\leq -1}\mathrm{i}^{k+l}\,H[k,l]_r\,z^{-r-1}\,, & |\lambda_2/\lambda_1| > |z|\,,\\ -(2\pi\mathrm{i})^{-1}\sum_{r\geq 0}\mathrm{i}^{k+l}\,H[k,l]_r\,z^{-r-1}\,, & |\lambda_2/\lambda_1| < |z|\,, \end{cases} \tag{6.33}$$

with coefficients being the loop algebra generators (6.17). We conclude that the soft graviton twistor representative acts as a generating function of the $Lw_{1+\infty}$ Hamiltonians $H[k,l]_r$.

Soft gravitons seem to be playing similar roles in the celestial and twistorial stories. Are these stories related to each other? Yes!

6.3 From Twistorial to Celestial $w_{1+\infty}$

We will now use twistor sigma models to make precise contact between the celestial and twistorial incarnations of $Lw_{1+\infty}$. To do this, recall the vertex operator (5.28) associated to a momentum eigenstate with twistor representative (6.26),

$$V(\omega,z,\bar{z}) \equiv V_h = \frac{1}{2\pi\mathrm{i}\hbar}\int_{\mathbb{P}^1}\frac{h|_Y\,\mathrm{D}\lambda}{\lambda_1^2\,\lambda_2^2}\,, \tag{6.34}$$

whose connected, tree-level correlators gave rise to MHV graviton amplitudes in Lemma 5.2. Here, we have introduced a formal loop counting parameter $\hbar$ that will be useful when truncating to 'tree-level' sigma model OPEs. Since the soft graviton representatives $h[k,l]$ only depend on the parameter $z \in \mathbb{P}^1$, we denote the soft graviton vertex operators by

$$V[k,l](z) \equiv V_{h[k,l]} = \frac{1}{2\pi\mathrm{i}\hbar}\int_{\mathbb{P}^1}\frac{h[k,l]|_Y\,\mathrm{D}\lambda}{\lambda_1^2\,\lambda_2^2} \tag{6.35}$$

We now postulate the dictionary

$$\begin{aligned} V(\omega, z, \bar{z}) &\longleftrightarrow \mathcal{O}_+(\omega, z, \bar{z})\,, \\ V[k,l](z) &\longleftrightarrow w[k,l](z) \end{aligned} \tag{6.36}$$

between operators in twistor sigma models and operators in CCFT. In Lemma 5.1, we explained how connected, tree-level correlators of $n-2$ vertex operators $V(\omega_i, z_i, \bar{z}_i)$ computed the n-graviton MHV tree-amplitude. Below, we show that the (tree-level) sigma model OPE of the operators on the left will reproduce the celestial OPE of those on the right. In this sense, twistor sigma models can potentially capture the self-dual (positive helicity) subsector of CCFT.

OPE of soft gravitons. Let us work in the affine patch $(\lambda_1, \lambda_2) = (1, \lambda)$. To streamline our OPE computations, we define new fields $\mu^{\dot\alpha}(\lambda)$ in our sigma models by the natural expressions

$$\mu^{\dot\alpha}(\lambda) = y^{\dot\alpha} + \lambda\,\tilde{y}^{\dot\alpha} + m^{\dot\alpha}(\lambda)\,, \tag{6.37}$$

where we are suppressing the space-time dependence of the fields. These are governed by the free theory given by the $h=0$ sigma model,

$$S_{h=0}[m] = -\frac{1}{4\pi \mathrm{i}\hbar}\int_{\mathbb{P}^1}\frac{\mathrm{d}\lambda\wedge\mathrm{d}\bar\lambda}{\lambda^2}\,[m\,\partial_{\bar\lambda}m]\,, \tag{6.38}$$

now written in our affine chart using $\mathrm{D}\lambda = \lambda_2\mathrm{d}\lambda_1 - \lambda_1\mathrm{d}\lambda_2 = -\mathrm{d}\lambda$. As a result, they obey the free-field OPE

$$\mu^{\dot\alpha}(\lambda)\,\mu^{\dot\beta}(\lambda') \sim 2\pi\mathrm{i}\hbar\,\frac{\lambda\,\lambda'\,\epsilon^{\dot\alpha\dot\beta}}{\lambda-\lambda'}\,, \tag{6.39}$$

with the factor of $\lambda\,\lambda'$ arising from the boundary conditions $m^{\dot\alpha}(\lambda) = 0$ at $\lambda = 0, \infty$.

The soft graviton vertex operators can be explicitly evaluated to yield

$$\begin{aligned} V[k,l](z) &= -\frac{1}{2\pi\mathrm{i}\hbar}\int_{\mathbb{P}^1}\frac{\mathrm{d}\lambda}{\lambda^2}\wedge\bar\delta(\lambda-z)\;\mathrm{i}^{k+l}\,\mu^{\dot 1}(\lambda)^k\,\mu^{\dot 2}(\lambda)^l \\ &= \frac{1}{2\pi\mathrm{i}\hbar}\,\frac{\mathrm{i}^{k+l-2}}{z^2}\,H[k,l](z)\,, \end{aligned} \tag{6.40}$$

where the $w_{1+\infty}$ Hamiltonian $H[k,l] = (\mu^{\dot 1})^k\,(\mu^{\dot 2})^l$ is now viewed as a composite field in our sigma model. Remarkably, the charges have turned into local operators in the sigma model. Since the delta function has set $\lambda = z$, one says that the sigma model Riemann sphere has been "pinned" to the celestial sphere.

Computing the OPE of two soft graviton vertex operators using (6.39), and retaining only the single contraction (i.e., tree-level) term, we find

$$V[k,l](z)\, V[m,n](w) \sim \frac{1}{2\pi \mathrm{i}\hbar} \frac{\mathrm{i}^{k+l+m+n-4}}{w^2} \frac{\{H[k,l], H[m,n]\}(w)}{z-w} + O(\hbar^0)\,, \tag{6.41}$$

having approximated $zw \sim w^2$. Applying the Poisson bracket relations (6.16) yields

$$V[k,l](z)\, V[m,n](w) \sim \frac{kn-lm}{z-w}\, V[k+m-1, l+n-1](w) + O(\hbar^0)\,. \tag{6.42}$$

As anticipated, to leading order in $\hbar$, this soft graviton sigma model OPE is isomorphic to the corresponding celestial OPE (6.11). Inserting this in tree-level sigma model correlators yields the corresponding collinear limits of positive helicity soft gravitons. This brings us full circle, proving that the $Lw_{1+\infty}$ symmetries observed in gravitational collinear limits and their dual celestial OPEs originate from the symmetries of SD GR.

OPE of hard gravitons. We can also derive the hard graviton celestial OPE (6.8) from the OPE of momentum eigenstate vertex operators.

Just as for a soft graviton, the integrals in the hard graviton vertex operator can be explicitly evaluated:

$$\begin{aligned} V(\omega, z, \bar{z}) &= -\frac{1}{2\pi \mathrm{i}\hbar} \int_{\mathbb{P}^1\times\mathbb{C}^*} \frac{\mathrm{d}\lambda}{\lambda^2}\frac{\mathrm{d}s}{s^3}\, \bar{\delta}(\sqrt{\omega}-s)\, \bar{\delta}(\sqrt{\omega}\, z - s\,\lambda)\, \mathrm{e}^{\mathrm{i}s\sqrt{\omega}[\mu(\lambda)\,\tilde{\zeta}]} \\ &= -\frac{1}{2\pi \mathrm{i}\hbar}\frac{1}{\omega^2 z^2} \exp\big[\mathrm{i}\omega\mu^{\dot{1}}(z) + \mathrm{i}\omega\bar{z}\mu^{\dot{2}}(z)\big]\,, \end{aligned} \tag{6.43}$$

where $\tilde{\zeta}_{\dot{\alpha}} = (1, \bar{z})$ as before. Computing the OPE by applying (6.39) gives us

$$\begin{aligned} &V(\omega_1, z_1, \bar{z}_1)\, V(\omega_2, z_2, \bar{z}_2) \\ &\sim \frac{1}{2\pi \mathrm{i}\hbar}\frac{1}{\omega_1^2\omega_2^2 z_2^2}\frac{1}{z_{12}} \big\{\exp\big[\mathrm{i}\omega_1\mu^{\dot{1}} + \mathrm{i}\omega_1\bar{z}_1\mu^{\dot{2}}\big], \exp\big[\mathrm{i}\omega_2\mu^{\dot{1}} + \mathrm{i}\omega_2\bar{z}_2\mu^{\dot{2}}\big]\big\}(z_2) \\ &\sim -\frac{1}{2\pi \mathrm{i}\hbar}\frac{1}{\omega_1\omega_2 z_2^2}\frac{\bar{z}_{12}}{z_{12}} \exp\big[\mathrm{i}(\omega_1+\omega_2)\mu^{\dot{1}}(z_2) + \mathrm{i}(\omega_1\bar{z}_1+\omega_2\bar{z}_2)\mu^{\dot{2}}(z_2)\big] \end{aligned} \tag{6.44}$$

to leading order in $\hbar$ and z_{12}. The right hand side is recognized to be another momentum eigenstate vertex operator,

$$V(\omega_1, z_1, \bar{z}_1)\, V(\omega_2, z_2, \bar{z}_2) \sim \frac{(\omega_1+\omega_2)^2}{\omega_1\omega_2}\frac{\bar{z}_{12}}{z_{12}}\, V\!\left(\omega_1+\omega_2, z_2, \frac{\omega_1\bar{z}_1+\omega_2\bar{z}_2}{\omega_1+\omega_2}\right). \tag{6.45}$$

Expanding this in small $\bar{z}_{12}$ reproduces (6.8). Equivalently, plugging (6.45) into sigma model correlators reproduces the collinear expansion (6.6) of graviton scattering.

With these calculations, we hope to have given the reader a flavor of the emerging connections between twistor theory and celestial holography. Admittedly, the investigations of this chapter are in early stages, leading to many possible avenues of future work. Some of the most pressing questions are the following:

- Can we use the ingredients of twistor sigma models to build a concrete example of celestial CFT for self-dual gravity?
- How do we move beyond self-duality and take negative helicity graviton operators into account?
- Does $Lw_{1+\infty}$ survive as a symmetry of full celestial CFT on going beyond the self-dual sector?

The current expectation is that an answer to the first question will emerge from studying holographic dualities for twistor string theories, wherein our twistor sigma models could act as building blocks of D1-brane worldvolume theories on $\mathbb{PT}$. A partial answer to the second question was provided in [37] in the form of a prescription to construct N^kMHV graviton amplitudes from sigma models governing higher degree rational curves in $\mathbb{PT}$. Similarly, a preliminary investigation of the third question was performed in [19] on the CCFT side, and in [38] on the twistor theory side.

References

1. Strominger A (2017) Lectures on the infrared structure of gravity and gauge theory. arXiv:1703.05448
2. Raclariu A-M (2021) Lectures on celestial holography. arXiv:2107.02075
3. Pasterski S (2021) Lectures on celestial amplitudes. arXiv:2108.04801
4. Aneesh PB, Compère G, de Gioia LP, Mol I, Swidler B (2021) Celestial holography: lectures on asymptotic symmetries. arXiv:2109.00997
5. Strominger A (2014) On BMS invariance of gravitational scattering. JHEP 07:152 [arXiv:1312.2229]
6. Strominger A (2014) Asymptotic symmetries of Yang-Mills theory. JHEP 07:151. arXiv:1308.0589
7. He T, Lysov V, Mitra P, Strominger A (2015) BMS supertranslations and Weinberg's soft graviton theorem. JHEP 05:151. arXiv:1401.7026
8. Adamo T, Casali E, Skinner D (2014) Perturbative gravity at null infinity. Class Quant Grav 31(22):225008. arXiv:1405.5122
9. Geyer Y, Lipstein AE, Mason L (2015) Ambitwistor strings at null infinity and (subleading) soft limits. Class Quant Grav 32(5):055003. arXiv:1406.1462
10. Adamo T, Casali E (2015) Perturbative gauge theory at null infinity. Phys Rev D 91(12):125022. arXiv:1504.02304
11. Pasterski S, Shao S-H, Strominger A (2017) Flat space amplitudes and conformal symmetry of the celestial sphere. Phys Rev D 96(6):065026. arXiv:1701.00049
12. Pasterski S, Shao S-H (2017) Conformal basis for flat space amplitudes. Phys Rev D 96(6):065022. arXiv:1705.01027
13. Donnay L, Puhm A, Strominger A (2019) Conformally soft photons and gravitons. JHEP 01:184. arXiv:1810.05219
14. Adamo T, Mason L, Sharma A (2019) Celestial amplitudes and conformal soft theorems. Class Quant Grav 36(20):205018. arXiv:1905.09224
15. Puhm A (2020) Conformally soft theorem in gravity. JHEP 09:130. arXiv:1905.09799
16. Guevara A (2019) Notes on conformal soft theorems and recursion relations in gravity. arXiv:1906.07810
17. Guevara A, Himwich E, Pate M, Strominger A (2021) Holographic symmetry algebras for gauge theory and gravity. arXiv:2103.03961
18. Strominger A (2021) w(1+infinity) and the celestial sphere. arXiv:2105.14346

19. Himwich E, Pate M, Singh K (2021) Celestial operator product expansions and $w_{1+\infty}$ symmetry for all spins. arXiv:2108.07763
20. Jiang H (2021) Holographic chiral algebra: supersymmetry, infinite ward identities, and EFTs. arXiv:2108.08799
21. Hoppe J (1989) Diffeomorphism groups, quantization and SU(infinity). Int J Mod Phys A 4:5235
22. Bakas I (1989) The large n limit of extended conformal symmetries. Phys Lett B 228:57
23. Boyer CP, Plebanski JF (1985) An infinite hierarchy of conservation laws and nonlinear superposition principles for self-dual einstein spaces. J Math Phys 26:229–234
24. Park Q-H (1990) Extended conformal symmetries in real heavens. Phys Lett B 236:429–432
25. Park Q-H (1990) Selfdual gravity as a large N limit of the two-dimensional nonlinear σ model. Phys Lett B 238:287–290
26. Mason LJ (1990) H-space: a universal integrable system? Twist Newsl 30:14–17
27. Newman ET (1976) Heaven and its properties. Gen Rel Grav 7:107–111
28. Penrose R (1976) The nonlinear graviton. Gen Rel Grav 7:171–176
29. Penrose R (1976) Nonlinear gravitons and curved twistor theory. Gen Rel Grav 7:31–52
30. He T, Mitra P, Strominger A (2016) 2D Kac-moody symmetry of 4D Yang-Mills theory. JHEP 10:137. arXiv:1503.02663
31. Kapec D, Mitra P, Raclariu A-M, Strominger A (2017) 2D stress tensor for 4D gravity. Phys Rev Lett 119(12):121601. arXiv:1609.00282
32. Fan W, Fotopoulos A, Taylor TR (2019) Soft limits of Yang-Mills amplitudes and conformal correlators. JHEP 05:121. arXiv:1903.01676
33. Pate M, Raclariu A-M, Strominger A, Yuan EY (2019) Celestial operator products of gluons and gravitons. arXiv:1910.07424
34. Bern Z, Dixon LJ, Perelstein M, Rozowsky JS (1999) Multileg one loop gravity amplitudes from gauge theory. Nucl Phys B 546:423–479 [hep-th/9811140]
35. Costello K, Paquette NM (2022) Celestial holography meets twisted holography: 4d amplitudes from chiral correlators. arXiv:2201.02595
36. Ward RS, Wells RO (1991) Twistor geometry and field theory. Cambridge Monographs on Mathematical Physics. Cambridge University Press, p 8
37. Adamo T, Mason L, Sharma A (2021) Twistor sigma models for quaternionic geometry and graviton scattering. arXiv:2103.16984
38. Adamo T, Bu W, Casali E, Sharma A (2021) Celestial operator products from the worldsheet. arXiv:2111.02279

Chapter 7
Twistors for SD Radiative Space-Times

Despite the perturbative non-linearity of the Einstein-Hilbert action [1–4], the scattering amplitudes of general relativity (GR) and supergravity (SUGRA) in Minkowski space-time are surprisingly simple. At tree-level, this is beautifully reflected in our running example of MHV amplitudes. In fact, there are now remarkably compact formulae for the *entire* tree-level S-matrix of GR in flat space which are totally divorced from the usual perspective of space-time Feynman rules [5, 6].

Given these remarkable achievements, it is natural to ask if analogous results can be found for gravitational scattering amplitudes in *curved* space-times. Of course, in a generic space-time the S-matrix will not exist due to particle creation effects by event horizons or singularities (cf., [7, 8]) or because the space-time is not asymptotically flat. But one can always consider scattering at large distances (i.e., probing weakly curved regions far from horizons) or the natural analogues of scattering amplitudes in asymptotically (anti-)de Sitter space-times (i.e., boundary correlation functions [9] or in-in correlators [10, 11]).

There are two broad reasons for being interested in such observables. The first is practical: they capture non-linear strong field effects which are invisible in Minkowski space and physically relevant (e.g., time-delay, the memory effect, tails, pair production). Their importance ranges from gravitational non-gaussianities in early universe cosmology [12, 13] to encoding infinite resummations of small momentum-transfer scattering in flat space [14–16]. The second reason is conceptual: scattering amplitudes in curved backgrounds are a theoretical playground where perturbative and non-perturbative physics meet. They test the robustness of on-shell techniques developed in Minkowski space, many of which simply break down in the presence of space-time curvature. For instance, in a generic space-time, momentum space Feynman rules and unitarity techniques fail, as Fourier transforms are obstructed by the background curvature and even tree-level scattering amplitudes are not rational functions of the kinematic data.

This tension between the importance and difficulty of computing scattering amplitudes in curved space-times means that, although they are often studied, nothing close

A. Sharma, *Twistor Sigma Models*, Springer Theses,
https://doi.org/10.1007/978-3-031-50751-9_7

to the all-multiplicity formulae available at tree-level in Minkowski space exists in even the simplest cases. For instance, state-of-the-art for tree-level graviton scattering in plane wave space-times is 3-points [17], while in AdS backgrounds—where CFT methods in the dual boundary theory can be used—it is 4-points (cf., [18–22]), with the exception of a particular R-symmetry sector of supergravity in $\text{AdS}_5 \times S^5$, where it is 5-points [23].[1] The gap between these low-point examples and the entire tree-level graviton S-matrix is profound and begs the question: is graviton scattering in curved space-time so complicated that we simply cannot hope to recover all-multiplicity results?

Over the course of this chapter and the next, we study tree-level graviton scattering in a large class of four-dimensional, (almost everywhere) asymptotically flat *self-dual* vacuum space-times. These space-times have a self-dual Weyl tensor determined by characteristic data which is a free function of three variables; as such, we refer to them as *self-dual radiative space-times*. The functional degrees of freedom in the curvature of such space-times means that there is no hope of obtaining high-multiplicity scattering amplitudes using standard space-time perturbation theory. But as we saw in Chap. 4, self-duality allows us to reformulate the perturbative expansion using twistor theory. Using twistor sigma models, we will derive an all-multiplicity expression for the tree-level MHV graviton amplitude on *any* self-dual radiative space-time. As an extension to the discussion in this thesis, we have also provided a conjecture for the tree-level graviton S-matrix in non-MHV helicity configurations on self-dual radiative space-times in [29].

7.1 SD Radiative Space-Times

A four-dimensional Lorentzian, vacuum space-time (M, g) is purely radiative if it is asymptotically flat with a smooth, topologically $\mathbb{R} \times S^2$ past or future null infinity ($\mathscr{I}^-$ or $\mathscr{I}^+$), and is completely determined by characteristic free data on $\mathscr{I}^\pm$ which obeys suitable regularity assumptions [30–32]. In other words, $\mathscr{I}^\pm$ is a good final/initial data surface for the Cauchy problem of radiative space-times. This characteristic data is composed of two functions at $\mathscr{I}^\pm$, encoded by the leading piece of the Weyl tensor in the asymptotic peeling expansion, which correspond to the self-dual and anti-self-dual degrees of freedom in the radiative gravitational field.

In this section, we review the definition of self-dual radiative space-times. They are obtained by complexifying the data and $\mathscr{I}^\pm$ and setting one of the characteristic functions to zero while keeping the other non-zero. Their twistor spaces can be constructed directly from the non-zero half of such characteristic data [33].

[1] There are some results in AdS which are not generic graviton boundary correlators but are nevertheless all-multiplicity. These include 'maximal U(1)-violating' correlators of components of the graviton multiplet in type IIB string theory on AdS_5 [24, 25], and integral kernels for the bulk dynamics of gravity in AdS_4 [26–28]. The latter do not explicitly encode boundary contributions to the AdS amplitudes and their relationship to position or momentum space expressions is obscure.

Homogeneous description of $\mathscr{I}^+$: An asymptotically flat space-time can be described in terms of retarded Bondi coordinates $(u, r, \zeta, \bar{\zeta})$, where u is retarded Bondi time, r is a parameter along the outgoing null geodesics of constant-u hypersurfaces and $(\zeta, \bar{\zeta})$ are stereographic coordinates on the celestial sphere. These coordinates are not global on the celestial sphere, so do not manifest Lorentz-invariance; this is remedied by instead working with homogeneous Bondi coordinates defined modulo the rescalings [34–39]:

$$(u, r, \lambda_\alpha, \bar{\lambda}_{\dot{\alpha}}) \sim (|b|^2 u,\ |b|^{-2} r,\ b\lambda_\alpha,\ \bar{b}\bar{\lambda}_{\dot{\alpha}}), \quad \forall\, b \in \mathbb{C}^* . \tag{7.1}$$

Here $\alpha, \dot{\alpha}$ are spinor indices of $\mathfrak{sl}(2, \mathbb{C})$, and the affine coordinates on the celestial sphere are simply recovered by going to a patch:

$$\lambda_\alpha = \frac{2^{1/4}}{\sqrt{1 + |\zeta|^2}}\,(1, \zeta)\,. \tag{7.2}$$

Choosing a time-like unit vector $t^{\alpha\dot{\alpha}}$ so that

$$y^{\alpha\dot{\alpha}} = \frac{u\, t^{\alpha\dot{\alpha}}}{\|\lambda\|^2} + r\, \lambda^\alpha\, \bar{\lambda}^{\dot{\alpha}}\,, \qquad t^{\alpha\dot{\alpha}} \lambda_\alpha \bar{\lambda}_{\dot{\alpha}} = \|\lambda\|^2 \equiv |\lambda_1|^2 + |\lambda_2|^2\,, \tag{7.3}$$

packages the Bondi coordinates in terms of spinorial coordinates $y^{\alpha\dot{\alpha}}$ on M. The choice of $t^{\alpha\dot{\alpha}}$ is equivalent to a choice of conformal factor $\|\lambda\|^2$ on the unit sphere (see also Sect. 2.1 of [40] for further discussion).

One advantage of this homogeneous formalism is that notions like spin- and conformal-weight can be formulated simply in terms of homogeneity under the projective rescalings (7.1) of the coordinates. There are natural C^∞ line bundles $\mathcal{O}(p, q)$ over space-time (or subsets thereof), whose sections transform as $f(|b|^2 u,\ |b|^{-2} r,\ b\lambda,\ \bar{b}\bar{\lambda}) = b^p \bar{b}^q\, f(u, r, \lambda, \bar{\lambda})$, and the usual notions of spin weight (s) and conformal weight (w) correspond to $s = \frac{p-q}{2}$ and $w = \frac{p+q}{2}$. In other words, spin/conformally-weighted functions correspond to functions valued in the line bundles $\mathcal{O}(p, q)$ in this formalism.

An asymptotically flat metric in Bondi-Sachs gauge admits a large-r expansion in these coordinates [41–44]:

$$\begin{aligned} ds^2 = \left(\frac{1}{\|\lambda\|^4} - 2\,\frac{m_B}{r}\right) du^2 + 2\, du\, dr - du\left(\eth\bar{\sigma}^0\, D\bar{\lambda} + \bar{\eth}\sigma^0\, D\lambda\right) \\ - r^2\left(D\lambda\, D\bar{\lambda} - \frac{\sigma^0}{r}\, D\lambda^2 - \frac{\bar{\sigma}^0}{r}\, D\bar{\lambda}^2\right) + O(r^{-2})\,, \end{aligned} \tag{7.4}$$

where $D\lambda := \langle\lambda\, d\lambda\rangle$, $D\bar{\lambda} := [\bar{\lambda}\, d\bar{\lambda}] = \overline{D\lambda}$, and $\eth, \bar{\eth}$ are the spin-weighted covariant derivatives on the sphere [35, 45],

$$\eth = \frac{\hat{\lambda}_\alpha}{\|\lambda\|^2}\frac{\partial}{\partial\lambda_\alpha}\,, \qquad \bar{\eth} = -\frac{\lambda_\alpha}{\|\lambda\|^2}\frac{\partial}{\partial\hat{\lambda}_\alpha}\,, \tag{7.5}$$

with $\hat{\lambda}_\alpha = t_\alpha{}^{\dot\alpha}\bar{\lambda}_{\dot\alpha} = (-\overline{\lambda_2}, \overline{\lambda_1})$ being the antipodal map that was previously referred to as quaternionic conjugation. The quantities $m_B, \boldsymbol{\sigma}^0, \bar{\boldsymbol{\sigma}}^0$ are spin- and conformally-weighted functions of $(u, \lambda, \bar{\lambda})$. In particular, for the line element to be homogeneous of degree zero, m_B must be valued in $\mathcal{O}(-3,-3)$, so is a spin-weight 0, conformal-weight -3 quantity; this is the Bondi mass aspect, which generalizes the Schwarzschild mass parameter for generic asymptotically flat space-times. Similarly, $\boldsymbol{\sigma}^0$ and $\bar{\boldsymbol{\sigma}}^0$ are valued in $\mathcal{O}(-3,1)$ and $\mathcal{O}(1,-3)$, so both have conformal weight -1 and spin-weights -2 and 2, respectively. These are the asymptotic shear optical scalars of the null hypersurfaces of constant u [41, 46, 47]. The (retarded) time evolution of the mass aspect m_B is controlled by $\boldsymbol{\sigma}^0$ (and its complex conjugate) through the asymptotic Bianchi identities (cf., [47, 48]).

Performing a conformal rescaling with conformal factor $R = r^{-1}$ gives

$$d\hat{s}^2 = -2\,du\,dR - D\lambda\, D\bar{\lambda} + R\,\boldsymbol{\sigma}^0 D\lambda^2 + R\,\bar{\boldsymbol{\sigma}}^0 D\bar{\lambda}^2 + O(R^4)\,. \tag{7.6}$$

The future null conformal boundary $\mathscr{I}^+$ is defined by $R \to 0$, and it is easy to see that $\mathscr{I}^+ \simeq \mathbb{R} \times S^2$ with degenerate conformal metric (cf., [49–52])

$$ds^2_{\mathscr{I}^+} = 0 \times du^2 + D\lambda\, D\bar{\lambda}\,. \tag{7.7}$$

The $(u, \lambda_\alpha, \bar{\lambda}_{\dot\alpha})$ provide homogeneous coordinates on $\mathscr{I}^+$ viewed as the total space of the line bundle $\mathcal{O}_{\mathbb{R}}(1,1) \to \mathbb{P}^1$, whose sections are real-valued functions obeying $f(b\lambda, \bar{b}\bar{\lambda}) = |b|^2 f(\lambda, \bar{\lambda})$. The BMS asymptotic symmetry group [42, 53] acts via Möbius transformations on the $\mathbb{P}^1$ base and supertranslations $u \to u + f$ for any f in $\mathcal{O}_{\mathbb{R}}(1,1)$, so there is no preferred choice of origin for the retarded time coordinate.[2]

Self-dual radiative metrics: The 2-spinor formalism decomposes the Weyl tensor C_{abcd} on M into self-dual (SD) and anti-self-dual (ASD) parts as [57]

$$C_{\alpha\dot\alpha\beta\dot\beta\gamma\dot\gamma\delta\dot\delta} = \tilde{\Psi}_{\dot\alpha\dot\beta\dot\gamma\dot\delta}\,\epsilon_{\alpha\beta}\,\epsilon_{\gamma\delta} + \Psi_{\alpha\beta\gamma\delta}\,\epsilon_{\dot\alpha\dot\beta}\,\epsilon_{\dot\gamma\dot\delta}\,, \tag{7.8}$$

with $\tilde{\Psi}_{\dot\alpha\dot\beta\dot\gamma\dot\delta}$ and $\Psi_{\alpha\beta\gamma\delta}$ being its SD and ASD Weyl spinors as seen in (3.7), (3.8). Near $\mathscr{I}^+$, the Weyl spinors exhibit the fall-offs

$$\Psi_{\alpha\beta\gamma\delta} = o_\alpha\, o_\beta\, o_\gamma\, o_\delta\, \frac{\psi_4}{r} + O(r^{-2})\,, \qquad \tilde{\Psi}_{\dot\alpha\dot\beta\dot\gamma\dot\delta} = \bar{o}_{\dot\alpha}\, \bar{o}_{\dot\beta}\, \bar{o}_{\dot\gamma}\, \bar{o}_{\dot\delta}\, \frac{\bar{\psi}_4}{r} + O(r^{-2})\,, \tag{7.9}$$

where $o_\alpha \bar{o}_{\dot\alpha} = \mathbf{e}_{\alpha\dot\alpha} u$, the outgoing null vector along constant u.

[2] This homogeneous framework is closely tied to descriptions of null infinity in terms of Carrollian geometry [54–56].

The coefficient functions $\psi_4(u, \lambda, \bar{\lambda})$, $\bar{\psi}_4(u, \lambda, \bar{\lambda})$ of this leading fall-off are known as the *radiation data*. They are explicitly given by (cf., [47, 58, 59]):

$$\psi_4 = -\partial_u^2 \boldsymbol{\sigma}^0 , \qquad \bar{\psi}_4 = -\partial_u^2 \bar{\boldsymbol{\sigma}}^0 , \tag{7.10}$$

with ψ_4 valued in $\mathcal{O}(-5, -1)$ and $\bar{\psi}_4$ valued in $\mathcal{O}(-1, -5)$. Thus, the complex shear $\boldsymbol{\sigma}^0$ serves as a key ingredient in any definition of the radiative phase space of asymptotically flat space-times [42, 60–66]. For instance, the *news function*

$$N(u, \lambda, \bar{\lambda}) := -\partial_u \bar{\boldsymbol{\sigma}}^0(u, \lambda, \bar{\lambda}) , \tag{7.11}$$

which takes values in $\mathcal{O}(0, -4)$ on $\mathscr{I}^+$, encodes the energy-momentum radiated through any interval of retarded time on $\mathscr{I}^+$ through the Bondi mass-loss theorem [42, 61].

To understand why $\boldsymbol{\sigma}^0$ defines the radiation data through ψ_4, it is instructive to first consider the case of a linear perturbation of complexified Minkowski space, $\mathbb{M} \equiv \mathbb{C}^4$. The whole of $\mathbb{M}$ is easily reconstructed from $\mathscr{I}^+$ in terms of light cone cuts: a point $x \in \mathbb{M}$ is described at $\mathscr{I}^+$ by the cut

$$u = x^{\alpha\dot{\alpha}} \lambda_\alpha \bar{\lambda}_{\dot{\alpha}} , \tag{7.12}$$

which is the spherical cross section corresponding to the intersection between $\mathscr{I}^+$ and the light cone with apex at x. This means that with suitable globality assumptions, it is easy to see that a linear gravitational perturbation on $\mathbb{M}$ is determined by the radiation data at $\mathscr{I}^+$ by means of the Kirchoff-d'Adhémar integral formulae adapted to null infinity [57, 59, 67].

In particular, to linear order, the ASD Weyl spinor is given by the spin-2 field

$$\psi_{\alpha\beta\gamma\delta}(x) = \int_{\mathbb{P}^1} \lambda_\alpha \lambda_\beta \lambda_\gamma \lambda_\delta \, \mathrm{D}\lambda \wedge \mathrm{D}\bar{\lambda} \left. \frac{\partial \psi_4}{\partial u} \right|_{u = \langle \lambda | x | \bar{\lambda}]} . \tag{7.13}$$

By differentiating under the integral sign, it is easily checked that this solves the linearized field equation $\partial^{\alpha\dot{\alpha}} \psi_{\alpha\beta\gamma\delta} = 0$ on flat space. In this formula, the integral is taken over the spherical cut of $\mathscr{I}^+$ corresponding to $x \in \mathbb{M}$ defined by (7.12). Conversely, the field defined by (7.13) is shown to give rise to the radiation data ψ_4 when the field is defined and differentiable over $\mathscr{I}^+$ including the vertex i^+ (i.e., future time-like infinity) [57].

For a generic asymptotically flat solution to the Einstein equations, the outgoing radiation field ψ_4 does not specify the full non-linear solution. Indeed, although the evolution of the mass aspect m_B in u is determined by the asymptotic Bianchi identities at $\mathscr{I}^+$, there is a constant of integration as $u \to \infty$ which encodes source contributions to the metric, such as a Schwarzschild mass. However, when such source terms are absent, ψ_4—and thus $\boldsymbol{\sigma}^0$—does provide good Cauchy data for the Einstein equations with $\mathscr{I}^+$ as the characteristic (final) data surface [30–32].

Such radiative space-times are therefore singled out by:

Definition 7.1 A *radiative* four-dimensional space-time is almost everywhere asymptotically flat and completely characterized by σ^0, so that $m_B \to 0$ at i^+, the future time-like infinity.

By 'almost everywhere asymptotically flat,' we mean space-times which are singular on isolated generators of $\mathscr{I}^+$ but otherwise admit a conformal compactification as usual. This allows important examples such as (multi-)plane wave space-times to fit into the radiative class.

In Lorentzian-real signature the SD and ASD parts of the Weyl tensor are related by complex conjugation; but for complexified metrics, or in Euclidean/split signature, they are independent. Thus, for a complexified radiative space-time, $\bar{\sigma}^0 \to \tilde{\sigma}^0$ which is independent free data no longer related to σ^0 by complex conjugation. The complexified data is viewed as living on a (partial) complexification of future null infinity, $\mathscr{I}^+_{\mathbb{C}} \simeq \mathbb{C} \times S^2$, for which Bondi time u is complex.

A complex vacuum space-time $M_{\mathbb{C}}$ is self-dual (SD) if $\Psi_{\alpha\beta\gamma\delta} = 0$ and anti-self-dual (ASD) if $\tilde{\Psi}_{\dot{\alpha}\dot{\beta}\dot{\gamma}\dot{\delta}} = 0$, and we can now define

Definition 7.2 A *self-dual (SD) radiative space-time* is a complex radiative space-time[3] determined by its radiative data $\tilde{\sigma}^0$, with $\sigma^0 = 0$.

Such a space-time is also referred to as an $\mathcal{H}$-space or "Heaven" (the space where good cones go!) [33, 69, 70]. They are tautologically holographic, with the 'bulk' metric determined entirely by the free data $\tilde{\sigma}^0$ on $\mathscr{I}^+_{\mathbb{C}}$.

Example: Self-dual plane wave metrics Several explicit SD radiative space-times have been constructed in the literature (e.g., [71, 72]), but a key example is provided by self-dual plane wave (SDPW) metrics. These are SD metrics with a five-dimensional Heisenberg algebra of Killing vectors, with centre given by a covariantly constant null vector n^a. Due to the necessary presence of such a null vector, such plane wave space-times do not admit Euclidean real slices. Fortunately, our computation of tree level scattering amplitudes will be agnostic to signature and these space-times will provide a running example throughout what follows.

An SDPW metric in spinorial coordinates $y^{\alpha\dot{\alpha}}$ takes the form

$$\mathrm{d}s^2 = \mathrm{d}y^{\alpha\dot{\alpha}}\,\mathrm{d}y_{\alpha\dot{\alpha}} + f(y^-)\,\mathrm{d}\tilde{z}^2\,, \tag{7.14}$$

where we have abbreviated $y^{2\dot{\alpha}} \equiv (\tilde{z}, y^-)$. In Lorentzian signature, y^- acts as a lightfront coordinate and $f(y^-)$ is a free function thereof. Let us also set $y^{1\dot{\alpha}} \equiv (y^+, z)$. Then $y^+, y^-, z, \tilde{z}$ provide Einstein-Rosen coordinates on our plane wave space-time.[4] In these coordinates, the null vector $n^a\partial_a = \partial_+$ is easily seen to be a covariantly constant Killing vector.

[3] For precise definitions of asymptotic flatness for complex radiative spacetimes, see [68].

[4] While Einstein-Rosen coordinates are not generally global due to null geodesic focusing in plane wave space-times [73], these SD examples do not suffer from this issue.

At this stage, we introduce a spinor dyad $o_\alpha = \tilde{o}_{\dot\alpha} = (1, 0)$, $\iota_\alpha = \tilde{\iota}_{\dot\alpha} = (0, 1)$, normalized so that $\langle \iota\, o \rangle = 1 = [\tilde\iota\, \tilde o]$. In terms of these, introduce a tetrad of 1-forms (i.e., a frame of the cotangent bundle):

$$\boldsymbol{\theta}^{\alpha\dot\alpha} = \mathrm{d}y^{\alpha\dot\alpha} + f(y^-)\, o^\alpha\, \tilde\iota^{\dot\alpha}\, \mathrm{d}\tilde z\,. \tag{7.15}$$

The dual tetrad (i.e., frame of the tangent bundle) is given by

$$\mathbf{e}_{\alpha\dot\alpha} = \frac{\partial}{\partial y_{\alpha\dot\alpha}} + f(y^-)\, o_\alpha\, \tilde\iota_{\dot\alpha}\, \frac{\partial}{\partial z}\,, \tag{7.16}$$

satisfying $\mathbf{e}_{\beta\dot\beta} \lrcorner\, \boldsymbol{\theta}^{\alpha\dot\alpha} = \delta^\alpha_\beta\, \delta^{\dot\alpha}_{\dot\beta}$. Dotted and undotted (equivalently, SD and ASD) spinor bundles can now be set up with respect to this tetrad.

To see that the metric (7.14) is indeed SD, one may compute the ASD and SD spin connections, finding

$$\tilde\Gamma_{\dot\alpha\dot\beta} = -\dot f\, \tilde\iota_{\dot\alpha}\, \tilde\iota_{\dot\beta}\, \mathrm{d}\tilde z\,, \qquad \Gamma_{\alpha\beta} = 0\,, \tag{7.17}$$

respectively, with $\dot f := f'(y^-)$. Thus the ASD part of the Riemann curvature 2-form is $R_{\alpha\beta} = \mathrm{d}\Gamma_{\alpha\beta} + \Gamma_\alpha{}^\gamma \wedge \Gamma_{\gamma\beta} = 0$, while the SD part is

$$\tilde R_{\dot\alpha\dot\beta} = \mathrm{d}\tilde\Gamma_{\dot\alpha\dot\beta} + \tilde\Gamma_{\dot\alpha}{}^{\dot\gamma} \wedge \tilde\Gamma_{\dot\gamma\dot\beta} = -\ddot f\, \tilde\iota_{\dot\alpha}\, \tilde\iota_{\dot\beta}\, \mathrm{d}y^- \wedge \mathrm{d}\tilde z\,, \tag{7.18}$$

so that the Weyl curvature of the metric is given by

$$\tilde\Psi_{\dot\alpha\dot\beta\dot\gamma\dot\delta} = -\ddot f\, \tilde\iota_{\dot\alpha}\, \tilde\iota_{\dot\beta}\, \tilde\iota_{\dot\gamma}\, \tilde\iota_{\dot\delta}\,, \qquad \Psi_{\alpha\beta\gamma\delta} = 0\,. \tag{7.19}$$

From now on, we also assume the sandwich condition on our SDPW metrics, meaning that $\ddot f$, and hence the curvature, is compactly supported in y^-. This ensures that the SDPW metric is asymptotically flat except in the n-direction and admits a well-defined S-matrix [17, 74].

As the SDPW metric is Kerr-Schild, its SD Weyl curvature tensor is essentially a linear field on the Minkowski background so the complex conjugate of (7.13) can be used to identify the radiation data giving rise to (7.19). The corresponding asymptotic data is distributional in nature:

$$\tilde{\boldsymbol{\sigma}}^0(u, \lambda, \bar\lambda)\, \mathrm{D}\bar\lambda = \frac{\langle o\, \lambda\rangle^2}{[\tilde o\, \bar\lambda]}\, \mathcal{F}\!\left(\frac{u}{\langle \lambda\, o\rangle\, [\bar\lambda\, \tilde o]}\right) \bar\delta\left(\langle \iota\, \lambda\rangle\right)\,, \tag{7.20}$$

with $\mathcal{F}(y^-) = \int^{y^-} f(s)\, \mathrm{d}s$ being an indefinite integral of f. This $\tilde{\boldsymbol{\sigma}}^0$ has distributional support along the generator of $\mathscr{I}^+$ corresponding to $(\lambda, \bar\lambda) = (\iota, \tilde\iota)$, and the associated news function is easily seen to be

$$N(u,\lambda,\bar{\lambda}) = \frac{\langle o\,\lambda\rangle}{[\tilde{o}\,\bar{\lambda}]^2}\, f\left(\frac{u}{\langle \lambda\, o\rangle\,[\bar{\lambda}\,\tilde{o}]}\right)\bar{\delta}(\langle \iota\,\lambda\rangle)\,. \tag{7.21}$$

7.2 Twistors from Heaven

For SD radiative space-times, we can build twistor representatives $\mathsf{h} \in \Omega^{0,1}(\mathbb{PT}, \mathcal{O}(2))$ satisfying the integrability condition

$$\mathsf{T} \equiv \bar{\partial}\mathsf{h} + \frac{1}{2}\{\mathsf{h},\mathsf{h}\} = 0 \tag{7.22}$$

directly in terms of the asymptotic radiative data [33, 35, 69]. To do so, we define a projection from the deformed twistor space to $\mathscr{I}^+_{\mathbb{C}}$:

$$p : \mathbb{P}\mathscr{T} \to \mathscr{I}^+_{\mathbb{C}}\,, \qquad (\mu^{\dot{\alpha}}, \lambda_\alpha) \mapsto (u = \mu^{\dot{\alpha}}\bar{\lambda}_{\dot{\alpha}}, \lambda_\alpha, \bar{\lambda}_{\dot{\alpha}})\,. \tag{7.23}$$

Then we can solve (7.22) by taking:

$$\mathsf{h} = \tilde{S}(u,\lambda,\bar{\lambda})\,\mathrm{D}\bar{\lambda}\,, \qquad \tilde{\boldsymbol{\sigma}}^0 = \partial_u \tilde{S}\,, \qquad \text{so} \quad \mathcal{L}_{\partial_{\dot{\alpha}}}\mathsf{h} = \bar{\lambda}_{\dot{\alpha}}\,\tilde{\boldsymbol{\sigma}}^0\,\mathrm{D}\bar{\lambda}\,. \tag{7.24}$$

Integrability (7.22) follows because h is holomorphic in $\mu^{\dot{\alpha}}$ and proportional to $\mathrm{D}\bar{\lambda}$. This allows us to define a deformed Dolbeault operator representing the complex structure deformation generated by h:

$$\overline{\nabla} :== \bar{\partial} + \{\mathsf{h}\,,\cdot\} = \bar{\partial} - \tilde{\boldsymbol{\sigma}}^0\,\mathrm{D}\bar{\lambda}\,\bar{\lambda}^{\dot{\alpha}}\,\partial_{\dot{\alpha}}\,. \tag{7.25}$$

It satisfies $\overline{\nabla}^2 = \{T,\cdot\} = 0$ and maps $(0,p)$-forms of $\mathbb{PT}$ into $(0,p)$-forms, thus providing a deformed differential on the complexes $\Omega^{0,\bullet}(\mathbb{PT},\mathcal{O}(n))$.

Let Y be a rational curve in $\mathbb{P}\mathscr{T}$ with equation $\mu^{\dot{\alpha}} = \mathsf{F}^{\dot{\alpha}}(y,\lambda)$. Using the relations $\mathrm{D}\bar{\lambda} = \mathrm{D}\hat{\lambda} = \langle\hat{\lambda}\,\lambda\rangle^2\,\bar{e}^0$ and $\bar{\partial}_0 \lrcorner\, \bar{e}^0 = 1$, the PDE (3.33) for holomorphic curves becomes

$$\bar{\partial}_0\mathsf{F}^{\dot{\alpha}} + \langle\hat{\lambda}\,\lambda\rangle^2\,\bar{\lambda}^{\dot{\alpha}}\,\tilde{\boldsymbol{\sigma}}^0 = 0\,, \tag{7.26}$$

Y can also be projected to $\mathscr{I}^+_{\mathbb{C}}$ by (7.23) to give a curved space analogue of a light cone cut (7.12), described by the cut function $u = \mathsf{Z}(y,\lambda) := \mathsf{F}^{\dot{\alpha}}(y,\lambda)\,\bar{\lambda}_{\dot{\alpha}}$. The Eq. (7.26) can then be written entirely in terms of data at $\mathscr{I}^+_{\mathbb{C}}$ and the "edth" operators (7.5),

$$\bar{\eth}^2\mathsf{Z} = \tilde{\boldsymbol{\sigma}}^0\,, \tag{7.27}$$

which is Newman's *good cut equation* written [33, 35, 69].

The moduli space of solutions of (7.26) or (7.27) has coordinates $y^{\alpha\dot{\alpha}}$ and acts as our complexified SD radiative space-time $M_{\mathbb{C}}$. If available, we can restrict to Euclidean real slices $M \subset M_{\mathbb{C}}$ by finding an involution on $\mathbb{P}\mathscr{T}$ as in the last point of Theorem 3.1. This imposes reality conditions on $y^{\alpha\dot{\alpha}}$. But as we are working perturbatively around flat space and complex analytically in $y^{\alpha\dot{\alpha}}$, we will be able to stay agnostic about these and work on any real slice M. This will be useful in our running example of (2, 2) signature SD plane wave backgrounds.

Reconstruction of the half-flat metric on M: Under our assumptions, standard theorems of Kodaira [75, 76] guarantee the existence of solutions $\mathsf{F}^{\dot{\alpha}}$ of (7.26)—or equivalently, of solutions Z to (7.27)—parametrized by $y \in M$. For the benefit of the reader, we now summarize the construction of the space-time metric from h.

To reconstruct the SD radiative metric on M from $\mathbb{P}\mathscr{T}$, observe that the vector field

$$\bar{\partial}_0 - \langle\hat{\lambda}\,\lambda\rangle^2\,\bar{\lambda}^{\dot{\alpha}}\,\tilde{\boldsymbol{\sigma}}^0\,\partial_{\dot{\alpha}}\,, \tag{7.28}$$

is anti-holomorphic in the deformed complex structure associated to $\overline{\nabla}$. This vector field annihilates the differential forms $e^{\dot{\alpha}} + \tilde{\boldsymbol{\sigma}}^0\,\mathrm{D}\bar{\lambda}\,\bar{\lambda}^{\dot{\alpha}}$, so these two 1-forms along with $\mathrm{D}\lambda$ span $\Omega^{1,0}(\mathbb{P}\mathscr{T})$. This allows us to construct a holomorphic 2-form $\Sigma \in \Omega^{2,0}(\mathbb{P}\mathscr{T}, \mathscr{O}(2))$ as

$$\Sigma = (e^{\dot{\alpha}} + \tilde{\boldsymbol{\sigma}}^0\,\bar{\lambda}^{\dot{\alpha}}\,\mathrm{D}\bar{\lambda}) \wedge (e_{\dot{\alpha}} + \tilde{\boldsymbol{\sigma}}^0\,\bar{\lambda}_{\dot{\alpha}}\,\mathrm{D}\bar{\lambda})\,. \tag{7.29}$$

As seen in the more general context of (3.23), $\mathrm{d}\Sigma = 0 \bmod \mathrm{D}\lambda$ because the complex structure deformation is hamiltonian. Following (3.36), we also see that such a closed 2-form is holomorphic along the curves Y for every $y \in M$. Liouville's theorem then implies that it gives rise to a basis of ASD 2-forms $\Sigma^{\alpha\beta}(y)$ over M.

Setting $\mu^{\dot{\alpha}} = \mathsf{F}^{\dot{\alpha}}(y, \lambda)$ and using (3.33), we obtain the pullback of Σ to $M \times \mathbb{P}^1$:

$$\begin{aligned} p^*\Sigma &= \mathrm{d}_y\mathsf{F}^{\dot{\alpha}} \wedge \mathrm{d}_y\mathsf{F}_{\dot{\alpha}} && \bmod \mathrm{D}\lambda \\ &\overset{!}{=} \lambda_\alpha\,\lambda_\beta\,\Sigma^{\alpha\beta}(y) && \bmod \mathrm{D}\lambda\,, \end{aligned} \tag{7.30}$$

where $\mathrm{d}_y = \mathrm{d}y^{\alpha\dot{\alpha}}\,\partial_{y^{\alpha\dot{\alpha}}}$ denotes the exterior derivative on $M \times \mathbb{P}^1$ along M. By construction, $\Sigma^{(\alpha\beta} \wedge \Sigma^{\gamma\delta)} = 0$. This implies the existence of a tetrad $\boldsymbol{\theta}^{\alpha\dot{\alpha}}$ on M for which [77]

$$\Sigma^{\alpha\beta} = \boldsymbol{\theta}^{\alpha\dot{\alpha}} \wedge \boldsymbol{\theta}^{\beta}{}_{\dot{\alpha}}\,. \tag{7.31}$$

The SD radiative metric on M is recovered as $\mathrm{d}s^2 = \boldsymbol{\theta}^{\alpha\dot{\alpha}}\,\boldsymbol{\theta}_{\alpha\dot{\alpha}}$. Self-duality and Ricci-flatness follow as a consequence of $\mathrm{d}\Sigma^{\alpha\beta} = 0$. To solve (7.30), the 1-form $\mathrm{d}_y\mathsf{F}^{\dot{\alpha}}$ must be of the form

$$\mathrm{d}_y\mathsf{F}^{\dot{\alpha}}(y, \lambda) = \mathsf{H}^{\dot{\alpha}}{}_{\dot{\beta}}(y, \lambda)\,\boldsymbol{\theta}^{\alpha\dot{\beta}}(y)\,\lambda_\alpha \tag{7.32}$$

with $\mathsf{H}^{\dot{\alpha}}{}_{\dot{\beta}}(y, \lambda) \in \mathrm{SL}(2, \mathbb{C})$. The latter provides a holomorphic trivialization of the dotted spinor bundle over the curve Y. The computation in (3.41) shows that it satisfies the PDE

$$\bar{\partial}_0\mathsf{H}^{\dot{\alpha}}{}_{\dot{\beta}} - N(\mathsf{Z}, \lambda, \bar{\lambda})\,\bar{\lambda}^{\dot{\alpha}}\,\bar{\lambda}_{\dot{\gamma}}\,\mathsf{H}^{\dot{\gamma}}{}_{\dot{\beta}} = 0 \tag{7.33}$$

on SD radiative backgrounds, where $N(\mathsf{Z}, \lambda, \bar{\lambda})$ is the (complexified) news tensor (7.11) evaluated on the good cut $u = \mathsf{Z}(y, \lambda, \bar{\lambda})$.

Due to (7.32), the SD vacuum equations admit a particularly clean Lax description on such space-times. Let $\mathbf{e}_{\alpha\dot{\alpha}}$ be the dual tetrad to $\theta^{\alpha\dot{\alpha}}$. Then (7.32) is equivalent to

$$\lambda_\alpha \, \mathsf{H}^{\dot{\alpha}}{}_{\dot{\beta}} = \mathbf{e}_{\alpha\dot{\beta}} \lrcorner \, \mathrm{d}_y \mathsf{F}^{\dot{\alpha}} = \mathbf{e}_{\alpha\dot{\beta}} \mathsf{F}^{\dot{\alpha}} \,, \tag{7.34}$$

which gives the Lax description

$$\lambda^\alpha \, \mathbf{e}_{\alpha\dot{\alpha}} \mathsf{F}^{\dot{\beta}} = 0 \,. \tag{7.35}$$

Integrability $[\lambda^\alpha \, \mathbf{e}_{\alpha\dot{\alpha}}, \lambda^\beta \, \mathbf{e}_{\beta\dot{\beta}}] = 0$ of the Lax pair $\lambda^\alpha \, \mathbf{e}_{\alpha\dot{\alpha}}$ is equivalent to the vacuum self-duality equations on M.

7.3 Gravitational Perturbations

We saw in Sect. 5.1 that metric perturbations $h_{ab}(y)$ on SD vacuum space-times M are called positive helicity if their ASD Weyl spinor perturbation vanishes: $\psi_{\alpha\beta\gamma\delta} = 0$. On the other hand, there is no invariant definition of negative helicity perturbations: these were defined by quotienting out the positive helicity perturbations from the space of all solutions of the linearized field equations.

As a consequence of the field equations and Bianchi identity, a negative helicity graviton can be characterized by $\mathbf{e}^{\alpha\dot{\alpha}} \psi_{\alpha\beta\gamma\delta} = 0$. This is the generalization of the free-field Eq. (2.27) to helicity -2 fields on space-times with zero ASD spin connection. The Penrose transform encodes such linearized fields on M through cohomology on $\mathbb{P}\mathscr{T}$ [78, 79]:

$$H^{0,1}(\mathbb{P}\mathscr{T}, \mathscr{O}(-6)) \simeq \left\{ h_{ab} \text{ on } M \,|\, \mathbf{e}^{\alpha\dot{\alpha}} \psi_{\alpha\beta\gamma\delta} = 0 \right\} \,. \tag{7.36}$$

When $\mathbb{P}\mathscr{T}$ is modeled by a deformation $\bar{\partial} \mapsto \overline{\nabla} = \bar{\partial} + \{\mathsf{h}, \cdot\}$ of the complex structure of $\mathbb{PT}$, this cohomology can be computed as the $\overline{\nabla}$-cohomology of the complex $\Omega^{0,\bullet}(\mathbb{PT}, \mathscr{O}(-6))$. Given some $\tilde{h} \in H^{0,1}(\mathbb{P}\mathscr{T}, \mathscr{O}(-6))$, the field on M is recovered from the integral formula

$$\psi_{\alpha\beta\gamma\delta}(y) = \int_{\mathsf{Y}} \mathrm{D}\lambda \wedge \lambda_\alpha \, \lambda_\beta \, \lambda_\gamma \, \lambda_\delta \, \tilde{h}|_{\mathsf{Y}} \,. \tag{7.37}$$

Such Penrose transforms are natural generalizations of their flat space cousins discussed in Sect. 2.3, developed in the general setting of arbitrary self-dual 4-manifolds in [78, 80].

On SD radiative space-times, we will be concerned with the special case when $\tilde{h}$ has no $\bar{e}^{\dot{\alpha}}$ components: $\tilde{h} = \tilde{h}_0 \, \bar{e}^0$. In particular, we can set $\tilde{h}|_{\mathsf{Y}} = \tilde{h}_0|_{\mathsf{Y}} \, \bar{e}^0$ in (7.37).

Since the background h given by (7.24) has no $\bar{e}^{\dot{\alpha}}$ component either, one can easily compute

$$\overline{\nabla}\tilde{h} = \bar{\partial}_{\dot{\alpha}}\tilde{h}_0\,\bar{e}^{\dot{\alpha}} \wedge \bar{e}^0 \stackrel{!}{=} 0 \implies \bar{\partial}_{\dot{\alpha}}\tilde{h}_0 = 0\,. \tag{7.38}$$

In complex coordinates, $\bar{\partial}_{\dot{\alpha}} = \langle\lambda\,\hat{\lambda}\rangle\,\partial/\partial\hat{\mu}^{\dot{\alpha}}$; so $\tilde{h}_0$ depends holomorphically on $\mu^{\dot{\alpha}}$. Then the Lax description (7.35) tells us that

$$\lambda^\alpha\,\mathbf{e}_{\alpha\dot{\alpha}}(\tilde{h}_0|_{\mathsf{Y}}) = \lambda^\alpha\,\mathbf{e}_{\alpha\dot{\alpha}}\mathsf{F}^{\dot{\beta}}\,(\partial_{\dot{\beta}}\tilde{h}_0)|_{\mathsf{Y}} = 0\,. \tag{7.39}$$

It follows that (7.37) solves the linear field equation $\mathbf{e}^{\alpha\dot{\alpha}}\psi_{\alpha\beta\gamma\delta} = 0$. The further requirement that $\psi_{\alpha\beta\gamma\delta}$ arises from an on-shell metric perturbation imposes extra conditions which generically obstruct setting $\tilde{\psi}_{\dot{\alpha}\dot{\beta}\dot{\gamma}\dot{\delta}} = 0$ due to the SD background curvature. That is, setting the linearized SD Weyl spinor to zero is inconsistent with diffeomorphism invariance (cf., [81]). We also remark that for Dolbeault representatives of the form $\tilde{h} = \tilde{h}_0\,\bar{e}^0$, (7.37) reduces to the Kirchoff-d'Adhémar formula (7.13) on flat space [82]. This is achieved with the identification $\partial_u\psi_4\,\mathrm{D}\bar{\lambda} = \tilde{h}$.

For positive helicity, the corresponding massless fields can only be constructed through a potential-modulo-gauge description. In particular, at spin 2 we can simply perturb the nonlinear graviton construction. The Penrose transform [78, 79] for a positive helicity graviton (with suitable analyticity) is

$$H^{0,1}(\mathbb{P}\mathscr{T}, \mathscr{O}(2)) \simeq \left\{h_{ab} \text{ on } M \mid \psi_{\alpha\beta\gamma\delta} = 0\right\}\,. \tag{7.40}$$

This can similarly be modeled by the $\overline{\nabla}$ cohomology of $\Omega^{0,\bullet}(\mathbb{PT}, \mathscr{O}(2))$.

For our purposes, again consider a representative $h = h_0\,\bar{e}^0 \in \Omega^{0,1}(\mathbb{PT}, \mathscr{O}(2))$. It is $\overline{\nabla}$-closed if and only if h_0 is holomorphic in $\mu^{\dot{\alpha}}$. On restriction to a rational curve Y, $h|_{\mathsf{Y}} \in H^{0,1}(\mathbb{P}^1, \mathscr{O}(2))$ because $\bar{\partial}|_{\mathsf{Y}}h|_{\mathsf{Y}} = \bar{\partial}_0(h_0|_{\mathsf{Y}})\,\bar{e}^0 \wedge \bar{e}^0 = 0$. But this cohomology group vanishes, so $h|_{\mathsf{Y}}$ must be $\bar{\partial}|_{\mathsf{Y}}$-exact:

$$h|_{\mathsf{Y}} = \bar{\partial}|_{\mathsf{Y}}j(y,\lambda)\,, \quad \text{i.e.,} \quad h_0|_{\mathsf{Y}} = \bar{\partial}_0 j(y,\lambda)\,, \tag{7.41}$$

for some function $j(y,\lambda)$ homogeneous of degree 2 in λ_α. Since h_0 was holomorphic in $\mu^{\dot{\alpha}}$, it follows that $\lambda^\alpha\mathbf{e}_{\alpha\dot{\alpha}}(h_0|_{\mathsf{Y}}) = 0$ using (7.35). Substituting for $h_0|_{\mathsf{Y}}$ from (7.41), this condition leads to $\bar{\partial}|_{\mathsf{Y}}(\lambda^\alpha\mathbf{e}_{\alpha\dot{\alpha}}j) = 0$. That is, $\lambda^\alpha\mathbf{e}_{\alpha\dot{\alpha}}j$ is globally holomorphic in λ_α, so

$$\lambda^\alpha\mathbf{e}_{\alpha\dot{\alpha}}j(y,\lambda) = \mathrm{i}\,\lambda^\alpha\lambda^\beta\lambda^\gamma\,\varphi_{\dot{\alpha}\alpha\beta\gamma}(y)\,, \tag{7.42}$$

for some function $\varphi_{\dot{\alpha}\alpha\beta\gamma}$ on M which is symmetric in its undotted indices. This is a potential for a metric perturbation h_{ab} [51]:

$$h_{\alpha\dot{\alpha}\beta\dot{\beta}} = \nabla^\gamma{}_{(\dot{\beta}}\varphi_{\dot{\alpha})\alpha\beta\gamma}\,, \tag{7.43}$$

where $\nabla = \mathbf{e} + \tilde{\Gamma}$ is the Levi-Civita connection on M. It is straightforward to confirm that the linearized ASD Weyl spinor $\psi_{\alpha\beta\gamma\delta}$ associated to the metric perturbation (7.43) vanishes, and thus h_{ab} defines a positive helicity graviton.

Momentum eigenstates: By virtue of their asymptotic flatness, we can consider graviton perturbations on SD radiative metrics which have asymptotic on-shell momentum $\kappa_\alpha \tilde{\kappa}_{\dot{\alpha}}$ at $\mathscr{I}_{\mathbb{C}}^-$. Although there is no notion of translation symmetry on a generic SD radiative background, we will continue to refer to such states as "momentum eigenstates". Explicit representatives for these can be built using twistor representatives that are functionally identical to those on Minkowski space, thanks to the form of the twistor complex structure for SD radiative space-times. For a positive helicity graviton, the representative

$$h = \int_{\mathbb{C}^*} \frac{\mathrm{d}s}{s^3}\, \bar{\delta}^2(\kappa - s\,\lambda)\, \mathrm{e}^{\mathrm{i}s[\mu\,\tilde{\kappa}]}\,, \tag{7.44}$$

is $\overline{\nabla}$-closed as it is a multiple of $\bar{e}^0$ and holomorphic in $\mu^{\dot{\alpha}}$. Restricting to a twistor curve $\mathsf{Y} \subset \mathbb{P}\mathscr{T}$ gives

$$h|_{\mathsf{Y}} = \frac{\langle \xi\,\lambda\rangle^3}{\langle \xi\,\kappa\rangle^3}\, \bar{\delta}(\langle \lambda\,\kappa\rangle)\, \mathrm{e}^{\mathrm{i}\phi} = \frac{1}{2\pi\mathrm{i}}\, \bar{\partial}|_{\mathsf{Y}} \left(\frac{\langle \xi\,\lambda\rangle^3}{\langle \xi\,\kappa\rangle^3\, \langle \lambda\,\kappa\rangle}\, \mathrm{e}^{\mathrm{i}\phi} \right), \tag{7.45}$$

where ξ_α is an arbitrary fixed reference spinor used to integrate out s in (7.44) against one of the delta functions, setting $s = \langle \xi\,\kappa\rangle / \langle \xi\,\lambda\rangle$. The function $\phi(y)$ is defined by

$$\phi(y) := [\mathsf{F}(y,\kappa)\,\tilde{\kappa}] = \mathsf{F}^{\dot{\alpha}}(y,\kappa)\,\tilde{\kappa}_{\dot{\alpha}}\,, \tag{7.46}$$

and it follows from (7.34) that this solves the Hamilton-Jacobi equation:

$$g^{ab}\, \partial_a \phi\, \partial_b \phi = 0\,, \tag{7.47}$$

with g^{ab} the inverse metric on M defined by the dual tetrad.

Comparing (7.45) with (7.41) yields

$$j(y,\lambda) = \frac{1}{2\pi\mathrm{i}}\, \frac{\langle \xi\,\lambda\rangle^3}{\langle \xi\,\kappa\rangle^3\, \langle \lambda\,\kappa\rangle}\, \mathrm{e}^{\mathrm{i}\phi}\,. \tag{7.48}$$

To descend to space-time via (7.42), define the background-dressed momentum $K_{\alpha\dot{\alpha}} := \mathbf{e}_{\alpha\dot{\alpha}}\phi$, for which

$$K_{\alpha\dot{\alpha}}(y) = \mathbf{e}_{\alpha\dot{\alpha}} \mathsf{F}^{\dot{\beta}}(y,\kappa)\, \tilde{\kappa}_{\dot{\beta}} = \kappa_\alpha\, \tilde{K}_{\dot{\alpha}}(y)\,, \qquad \text{where } \tilde{K}_{\dot{\alpha}}(y) := \tilde{\kappa}_{\dot{\beta}}\, \mathsf{H}^{\dot{\beta}}{}_{\dot{\alpha}}(y,\kappa)\,, \tag{7.49}$$

having used (7.34). In other words, the holomorphic frame $\mathsf{H}^{\dot{\alpha}}{}_{\dot{\beta}}$ serves to dress the dotted components of the graviton momentum as it traverses the SD radiative background. The potential on M associated with (7.48) is easily seen to be

$$\varphi_{\dot{\alpha}\alpha\beta\gamma} = \mathrm{i}\,\frac{\xi_\alpha\,\xi_\beta\,\xi_\gamma\,\tilde{K}_{\dot{\alpha}}}{\langle\xi\,\kappa\rangle^3}\,\mathrm{e}^{\mathrm{i}\phi}\,, \tag{7.50}$$

from which the positive helicity graviton perturbation follows via (7.43):

$$h^{(+)}_{\alpha\dot{\alpha}\beta\dot{\beta}} = \mathbf{e}^{\gamma}{}_{(\dot{\alpha}}\varphi_{\dot{\beta})\alpha\beta\gamma} - \tilde{\Gamma}^{\dot{\gamma}}{}_{(\dot{\alpha}}{}^{\gamma}{}_{\dot{\beta})}\,\varphi_{\dot{\gamma}\alpha\beta\gamma}\,, \tag{7.51}$$

where $\tilde{\Gamma}_{\dot{\alpha}\dot{\beta}} \equiv \tilde{\Gamma}_{\dot{\alpha}\dot{\beta}\gamma\dot{\gamma}}\,\boldsymbol{\theta}^{\gamma\dot{\gamma}}$ is the background SD spin connection

For a negative helicity graviton, the representative

$$\tilde{h} = \int_{\mathbb{C}^*} \mathrm{d}s\, s^5\, \bar{\delta}^2(\kappa - s\,\lambda)\,\mathrm{e}^{\mathrm{i}s[\mu\,\tilde{\kappa}]}\,, \tag{7.52}$$

is $\overline{\nabla}$-closed and valued in $\mathscr{O}(-6)$. Feeding this into (7.37) immediately gives

$$\psi_{\alpha\beta\gamma\delta} = \kappa_\alpha\,\kappa_\beta\,\kappa_\gamma\,\kappa_\delta\,\mathrm{e}^{\mathrm{i}\phi}\,, \tag{7.53}$$

which is easily seen to obey $\mathbf{e}^{\alpha\dot{\alpha}}\psi_{\alpha\beta\gamma\delta} = 0$. Note that this differs from the expression for the linearized Weyl curvature of a negative helicity graviton on Minkowski space only by the substitution $\mathrm{e}^{\mathrm{i}k\cdot y} \to \mathrm{e}^{\mathrm{i}\phi(y)}$; this is an expected consequence of the self-duality of the background.

Example: SDPWs The twistor construction of SDPW metrics was one of the first examples of the non-linear graviton construction [83–85]; our presentation differs from those in the literature by making use of a Dolbeault (rather than Čech) framework. The complex structure $\overline{\nabla} = \bar{\partial} + \{\mathsf{h}, \cdot\}$ is defined by

$$\mathsf{h} = 2\pi\mathrm{i}\int_{\mathbb{C}^*} \frac{\mathrm{d}s}{s^3}\,\bar{\delta}^2(\iota - s\,\lambda)\,\mathscr{F}(s[\mu\,\tilde{\iota}])\,, \qquad \text{where } \mathscr{F}(y^-) := \int^{y^-} \mathcal{F}(t)\,\mathrm{d}t\,, \tag{7.54}$$

and $\mathcal{F}(y^-) = \int^{y^-} f(t)\,\mathrm{d}t$ is the first antiderivative of $f(y^-)$. Observing that

$$\mathcal{L}_{\partial_{\dot{\alpha}}}\mathsf{h}\big|_{\mathsf{Y}} = \bar{\partial}|_{\mathsf{Y}}\left(\frac{\langle o\,\lambda\rangle^2}{\langle\iota\,\lambda\rangle}\,\tilde{\iota}^{\dot{\alpha}}\,\mathcal{F}(y^-)\right)\,, \tag{7.55}$$

it follows from (3.33) that the holomorphic curves in $\mathbb{P}\mathscr{T}$ are given by

$$\mathsf{Y}:\quad \mu^{\dot{\alpha}} = \mathsf{F}^{\dot{\alpha}}(y,\lambda) = y^{\alpha\dot{\alpha}}\lambda_\alpha + \frac{\langle o\,\lambda\rangle^2}{\langle\iota\,\lambda\rangle}\,\tilde{\iota}^{\dot{\alpha}}\,\mathcal{F}(y^-)\,. \tag{7.56}$$

With the SDPW tetrad (7.15), Eq. (7.32) produces the associated holomorphic frame $\mathsf{H}^{\dot{\alpha}}{}_{\dot{\beta}}$ for the SD spin bundle:

$$\mathsf{H}^{\dot{\alpha}}{}_{\dot{\beta}}(y,\lambda) = \delta^{\dot{\alpha}}_{\dot{\beta}} - \frac{\langle o\,\lambda\rangle}{\langle\iota\,\lambda\rangle}\,f(y^-)\,\tilde{\iota}^{\dot{\alpha}}\,\tilde{\iota}_{\dot{\beta}}\,. \tag{7.57}$$

It is easy to confirm that this $\mathsf{H}^{\dot\alpha}{}_{\dot\beta}$ satisfies (7.34).

The solution to the Hamilton-Jacobi equations on a SDPW metric associated with incoming asymptotic momentum $\kappa_\alpha\tilde\kappa_{\dot\alpha}$ is thus

$$\phi(y) = k\cdot y + \frac{k^2}{k_+}\,\mathcal{F}(y^-) = \langle\kappa|y|\tilde\kappa] + \frac{\langle o\,\kappa\rangle^2\,[\tilde\iota\,\tilde\kappa]}{\langle\iota\,\kappa\rangle}\,\mathcal{F}(y^-)\,, \tag{7.58}$$

with the momenta parametrized in lightfront coordinates $k_\mu = (k_-, k, \tilde k, k_+)$. In fact, this is the SD truncation of the solution to the Hamilton-Jacobi equation on a general plane wave space-time [86, 87]. The null Killing vector $n^{\alpha\dot\alpha} = \iota^\alpha\tilde\iota^{\dot\alpha}$ of the SDPW metric enables us to impose lightfront gauge on graviton perturbations of both helicities. Applying (7.57) to (7.51) gives the positive helicity graviton

$$h^{(+)}_{\alpha\dot\alpha\beta\dot\beta} = \frac{\xi_\alpha\,\xi_\beta}{\langle\xi\,\kappa\rangle^2}\left(\tilde K_{\dot\alpha}\,\tilde K_{\dot\beta} - \mathrm{i}\,\dot f\,\frac{[\tilde\iota\,\tilde\kappa]}{\langle\iota\,\kappa\rangle}\,\tilde\iota_{\dot\alpha}\,\tilde\iota_{\dot\beta}\right)\mathrm{e}^{\mathrm{i}\phi}\,, \tag{7.59}$$

where choosing the reference spinor $\xi_\alpha = \iota_\alpha$ imposes lightfront gauge, and the dressed graviton momentum is

$$K_{\alpha\dot\alpha}(y^-) = \kappa_\alpha\,\tilde K_{\dot\alpha}(y^-)\,, \qquad \tilde K_{\dot\alpha} = \tilde\kappa_{\dot\alpha} - \frac{\langle o\,\kappa\rangle\,[\tilde\iota\,\tilde\kappa]\,f(y^-)}{\langle\iota\,\kappa\rangle}\,\tilde\iota_{\dot\alpha}\,. \tag{7.60}$$

The explicit 'tail' term in $h^{(+)}_{ab}$, proportional to $\dot f$, is a result of the failure of Huygens' principle in the non-linear SD background [88].

For negative helicity gravitons, the lightfront gauge condition is precisely what is required to consistently set the associated linearized SD Weyl spinor $\tilde\psi_{\dot\alpha\dot\beta\dot\gamma\dot\delta}$ to zero. In particular, the negative helicity graviton associated with (7.53) is given by

$$h^{(-)}_{\alpha\dot\alpha\beta\dot\beta} = \frac{\kappa_\alpha\,\kappa_\beta}{[\tilde\iota\,\tilde\kappa]^2}\,\tilde\iota_{\dot\alpha}\,\tilde\iota_{\dot\beta}\,\mathrm{e}^{\mathrm{i}\phi}\,. \tag{7.61}$$

The combined de Donder-lightfront gauge is the only one in which it is possible to make this identification of the negative helicity graviton at metric level.

Having computed the twistor spaces of SD radiative space-times and twistor representatives of graviton states thereon, we finally come to computing some amplitudes on curved space-times!

References

1. DeWitt BS (1967) Quantum theory of gravity. 1. The canonical theory. Phys Rev 160:1113–1148
2. DeWitt BS (1967) Quantum theory of gravity. 2. The manifestly covariant theory. Phys Rev 162:1195–1239

3. DeWitt BS (1967) Quantum theory of gravity. 3. Applications of the covariant theory. Phys Rev 162:1239–1256
4. 't Hooft G, Veltman MJG (1974) One loop divergencies in the theory of gravitation. Ann Inst H Poincare Phys Theor A 20:69–94
5. Cachazo F, Skinner D (2013) Gravity from rational curves in twistor space. Phys Rev Lett 110(16):161301. arXiv:1207.0741
6. Cachazo F, He S, Yuan EY (2014) Scattering of massless particles in arbitrary dimensions. Phys Rev Lett 113(17):171601. arXiv:1307.2199
7. DeWitt BS (1975) Quantum field theory in curved space-time. Phys Rept 19:295–357
8. Birrell ND, Davies PCW (1984) Quantum fields in curved space. Cambridge monographs on mathematical physics. Cambridge University Press, Cambridge, UK, vol 2
9. Witten E (1998) Anti-de sitter space and holography. Adv Theor Math Phys 2:253–291 hep-th/9802150
10. Witten E (2001) Quantum gravity in de Sitter space. In: Strings 2001: international conference, vol 6. arXiv:hep-th/0106109
11. Strominger A (2001) The dS/CFT correspondence. JHEP 10:034 hep-th/0106113
12. Mukhanov VF, Feldman HA, Brandenberger RH (1992) Theory of cosmological perturbations. Phys Rept 215:203–333
13. Maldacena JM (2003) Non-Gaussian features of primordial fluctuations in single field inflationary models. JHEP 05:013 astro-ph/0210603
14. 't Hooft G (1987) Graviton dominance in ultrahigh-energy scattering. Phys Lett B 198:61–63
15. Kabat DN, Ortiz M (1992) Eikonal quantum gravity and Planckian scattering. Nucl Phys B 388:570–592 hep-th/9203082
16. Adamo T, Cristofoli A, Tourkine P (2021) Eikonal amplitudes from curved backgrounds. arXiv:2112.09113
17. Adamo T, Casali E, Mason L, Nekovar S (2018) Scattering on plane waves and the double copy. Class Quant Grav 35(1):05004. arXiv:1706.08925
18. D'Hoker E, Freedman DZ, Mathur SD, Matusis A, Rastelli L (1999) Graviton exchange and complete four point functions in the AdS/CFT correspondence. Nucl Phys B 562:353–394 hep-th/9903196
19. Raju S (2012) Four point functions of the stress tensor and conserved currents in AdS_4/CFT_3. Phys Rev D 85:126008 arXiv:1201.6452
20. Rastelli L, Zhou X (2017) Mellin amplitudes for $AdS_5 \times S^5$. Phys Rev Lett 118(9):091602. arXiv:1608.06624
21. Caron-Huot S, Trinh A-K (2019) All tree-level correlators in $AdS_5 \times S_5$ supergravity: hidden ten-dimensional conformal symmetry. JHEP 01:196 arXiv:1809.09173
22. Alday LF, Zhou X (2020) All holographic four-point functions in all maximally supersymmetric CFTs. arXiv:2006.12505
23. Gonçalves V, Pereira R, Zhou X (2019) $20'$ five-point function from $AdS_5 \times S^5$ supergravity. JHEP 10:247 arXiv:1906.05305
24. Green MB, Wen C (2021) Maximal $U(1)_Y$-violating n-point correlators in $\mathcal{N} = 4$ super-Yang-Mills theory. JHEP 02:042 arXiv:2009.01211
25. Dorigoni D, Green MB, Wen C (2021) Exact expressions for n-point maximal $U(1)_Y$-violating integrated correlators in $SU(N)$ $\mathcal{N} = 4$ SYM. JHEP 11:132 arXiv:2109.08086
26. Adamo T, Mason L (2014) Conformal and Einstein gravity from twistor actions. Class Quant Grav 31(4):045014. arXiv:1307.5043
27. Adamo T (2015) Gravity with a cosmological constant from rational curves. JHEP 11:098 arXiv:1508.02554
28. Adamo T, Mason L, Sharma A (2021) Twistor sigma models for quaternionic geometry and graviton scattering. arXiv:2103.16984
29. Adamo T, Mason L, Sharma A (2022) Graviton scattering in self-dual radiative space-times. arXiv:2203.02238
30. Friedrich H (1986) On purely radiative space-times. Commun Math Phys 103:35–65

31. Friedrich H (2013) The Taylor expansion at past time-like infinity. Commun Math Phys 324:263–300 arXiv:1306.5626
32. Chruściel PT, Paetz T-T (2013) Solutions of the vacuum Einstein equations with initial data on past null infinity. Class Quant Grav 30:235037 arXiv:1307.0321
33. Newman ET (1976) Heaven and its properties. Gen Rel Grav 7:107–111
34. Sparling GAJ (1990) Dynamically broken symmetry and global Yang-Mills in Minkowski Space. In: Mason LJ, Hughston LP (eds) Further advances in twistor theory, vol 231, chap. 1.4.2. Pitman research notes in mathematics
35. Eastwood M, Tod P (1982) Edth—a differential operator on the sphere. Math Proc Camb Phil Soc 92:317–330
36. Adamo T, Casali E, Skinner D (2014) Perturbative gravity at null infinity. Class Quant Grav 31(22):225008. arXiv:1405.5122
37. Geyer Y, Lipstein AE, Mason L (2015) Ambitwistor strings at null infinity and (subleading) soft limits. Class Quant Grav 32(5):055003. arXiv:1406.1462
38. Adamo T, Casali E (2015) Perturbative gauge theory at null infinity. Phys Rev D91(12):125022. arXiv:1504.02304
39. Adamo T, Kol U (2021) Classical double copy at null infinity. arXiv:2109.07832
40. Adamo T, Mason L, Sharma A (2020) Gluon scattering on self-dual radiative gauge fields. arXiv:2010.14996
41. Sachs RK (1961) Gravitational waves in general relativity 6. The outgoing radiation condition. Proc Roy Soc Lond A264:309–338
42. Bondi H, van der Burg MGJ, Metzner AWK (1962) Gravitational waves in general relativity 7. Waves from axisymmetric isolated systems. Proc Roy Soc Lond A269:21–52
43. Sachs RK (1962) Gravitational waves in general relativity 8. Waves in asymptotically flat space-times. Proc Roy Soc Lond A270:103–126
44. Mädler T, Winicour J (2016) Bondi-Sachs formalism. Scholarpedia 11:33528 arXiv:1609.01731
45. Goldberg JN, MacFarlane AJ, Newman ET, Rohrlich F, Sudarshan ECG (1967) Spin-s spherical harmonics and Edth. J Math Phys 8:2155
46. Jordan P, Ehlers J, Sachs RK (1961) Beiträge zur Theorie der reinen Gravitationsstrahlung. Akad Wiss Lit Mainz, Abh Math Nat Kl 1:1–61
47. Newman E, Penrose R (1962) An Approach to gravitational radiation by a method of spin coefficients. J Math Phys 3:566–578
48. Adamo TM, Kozameh CN, Newman ET (2009) Null geodesic congruences, asymptotically flat space-times and their physical interpretation. Living Rev Rel 12:6 arXiv:0906.2155
49. Penrose R (1963) Asymptotic properties of fields and space-times. Phys Rev Lett 10:66–68
50. Penrose R (1964) Conformal treatment of infinity. In: DeWitt C, DeWitt B (eds) Relativity, groups and topology. Gordon and Breach, pp 565–586
51. Penrose R (1965) Zero rest mass fields including gravitation: asymptotic behavior. Proc Roy Soc Lond A284:159
52. Frauendiener J (2004) Conformal infinity. Living Rev Rel 7:1
53. Sachs R (1962) Asymptotic symmetries in gravitational theory. Phys Rev 128:2851–2864
54. Figueroa-O'Farrill J, Have E, Prohazka S, Salzer J (2021) Carrollian and celestial spaces at infinity. arXiv:2112.03319
55. Herfray Y (2021) Carrollian manifolds and null infinity: a view from Cartan geometry. arXiv:2112.09048
56. Donnay L, Fiorucci A, Herfray Y, Ruzziconi R (2022) A carrollian perspective on celestial holography. arXiv:2202.04702
57. Penrose R, Rindler W (1984) Spinors and space-time, vol 1. Cambridge monographs on mathematical physics. Cambridge University Press, Cambridge, UK
58. Newman ET, Tod KP (1981) Asymptotically flat space-times. In: Held A (ed) General relativity and gravitation: one hundred years after the birth of Albert Einstein, vol 2. Plenum Press, pp 1–36

59. Penrose R, Rindler W (1986) Spinors and space-time, vol 2. Cambridge monographs on mathematical physics. Cambridge University Press, Cambridge, UK
60. Sachs RK (1962) On the characteristic initial value problem in gravitational theory. J Math Phys 3:908–914
61. Newman ET, Unti TWJ (1962) Behavior of asymptotically flat empty spaces. J Math Phys 3(5):891
62. Ashtekar A, Streubel M (1981) Symplectic geometry of radiative modes and conserved quantities at null infinity. Proc Roy Soc Lond A 376:585–607
63. Ashtekar A (1981) Radiative degrees of freedom of the gravitational field in exact general relativity. J Math Phys 22:2885–2895
64. Ashtekar A (1987) Asymptotic quantization: based on 1984 Naples lectures. Bibliopolis
65. Herfray Y (2020) Asymptotic shear and the intrinsic geometry of null-infinity. J Math Phys 61(7):072502. arXiv:2001.01281
66. Herfray Y (2021) Tractor geometry of asymptotically flat space-times. arXiv:2103.10405
67. Penrose R (1980) Null hypersurface initial data for classical fields of arbitrary spin and for general relativity. Gen Rel Grav 12:225–264
68. Ludvigsen M, Newman ET, Tod KP (1981) Asymptotically flat H-spaces. J Math Phys 22:818–823
69. Hansen RO, Newman ET, Penrose R, Tod KP (1978) The metric and curvature properties of H-space. Proc Roy Soc Lond A 363:445–468
70. Ko M, Ludvigsen M, Newman ET, Tod KP (1981) The theory of H-space. Phys Rep 71(2):51–139
71. Tod KP, Ward RS (1979) Selfdual metrics with selfdual Killing vectors. Proc Roy Soc Lond A 368:411–427
72. Sparling GAJ, Tod KP (1981) An example of an H-space. J Math Phys 22:331–332
73. Penrose R (1965) A remarkable property of plane waves in general relativity. Rev Mod Phys 37:215–220
74. Gibbons GW (1975) Quantized fields propagating in plane wave space-times. Commun Math Phys 45:191–202
75. Kodaira K (1962) A theorem of completeness of characteristic systems for analytic families of compact submanifolds of complex manifolds. Ann Math 75:146–162
76. Kodaira K (1963) On stability of compact submanifolds of complex manifolds. Am J Math 85:79–94
77. Capovilla R, Jacobson T, Dell J, Mason LJ (1991) Selfdual two forms and gravity. Class Quant Grav 8:41–57
78. Hitchin NJ (1980) Linear field equations on selfdual spaces. Proc Roy Soc Lond A 370:173–191
79. Ward RS, Wells RO (1991) Twistor geometry and field theory. Cambridge monographs on mathematical physics, vol 8. Cambridge University Press
80. Woodhouse NMJ (1985) Real methods in twistor theory. Class Quant Grav 2:257–291
81. Mason LJ, Skinner D (2010) Gravity, twistors and the MHV formalism. Commun Math Phys 294:827–862 arXiv:0808.3907
82. Mason LJ (1990) Dolbeault representative from characteristic initial data at null infinity. In: Mason LJ, Hughston LP (eds) Further advances in twistor theory, vol. 231, chap. 1.2.16. Pitman Research Notes in Mathematics
83. Ward RS (1978) A class of self-dual solutions of Einstein's equations. Proc Roy Soc Lond A 363(1713):289–295
84. Curtis WD, Miller FR, Lerner DE (1978) Complex pp waves and the nonlinear graviton construction. J Math Phys 19:2024–2027
85. Porter JR (1982) The nonlinear graviton: superposition of plane waves. Class Quantum Gravity 14:1023–1033
86. Ward RS (1987) Progressing waves in flat space-time and in plane wave space-times. Class Quantum Gravity 4:775–778
87. Adamo T, Ilderton A (2020) Classical and quantum double copy of back-reaction. JHEP 09:200 arXiv:2005.05807
88. Harte AI (2013) Tails of plane wave spacetimes: wave-wave scattering in general relativity. Phys Rev D88(8):084059. arXiv:1309.5020

Chapter 8
Graviton Scattering in SD Radiative Space-Times

Our description of the non-linear graviton construction in Theorem 3.1 was perturbative in nature. We computed the curved twistor space $\mathbb{P}\mathscr{T}$ associated to an SD vacuum space-time M as a deformation of 'flat' twistor space $\mathbb{PT} = \mathscr{O}(1) \oplus \mathscr{O}(1)$. In this sense, $\mathbb{PT}$ acted as our background twistor space. We could instead decide to use a given curved twistor space $\mathbb{P}\mathscr{T}$ with representative h as our background. This will allow us to study SD vacuum space-times $\mathcal{M}$ obtained from perturbing the metric of M as perturbations $\mathsf{h} \mapsto \mathsf{h} + h$, with $h \in \Omega^{0,1}(\mathbb{PT}, \mathscr{O}(2))$ satisfying the background-coupled integrability equation

$$T = \overline{\nabla} h + \frac{1}{2}\{h, h\} = 0\,. \tag{8.1}$$

This is obtained by perturbing the integrability condition $\mathsf{T} \equiv \bar{\partial}\mathsf{h} + \frac{1}{2}\{\mathsf{h}, \mathsf{h}\} = 0$ for the background, and expressing anti-holomorphic derivatives in terms of the deformed Dolbeault operator $\overline{\nabla} h = \bar{\partial} h + \{\mathsf{h}, h\}$.

In this chapter, we will construct hyperkähler metrics on perturbations $\mathcal{M}$ of SD radiative space-times, starting from data h satisfying $T = 0$. To do this, we will explain how to expand our twistor sigma models around the background twistor representative h of an SD radiative space-time. Equations of motion of the resulting models govern deformations of holomorphic curves in the associated background twistor space $\mathbb{P}\mathscr{T}$. In this sense, we will call them background-coupled twistor sigma models. Subsequently, these will be used to compute tree-level graviton MHV amplitudes on SD radiative space-times. We will also illustrate the resulting formulae in the example of SD plane wave space-times.

A. Sharma, *Twistor Sigma Models*, Springer Theses,
https://doi.org/10.1007/978-3-031-50751-9_8

8.1 Background-Coupled Sigma Models

Let M possess an SD vacuum metric g_0 with tetrad $\theta_0^{\alpha\dot\alpha}$. Let $\mathbb{P}\mathscr{T}$ denote the corresponding twistor space. As usual, we will stay agnostic about space-time signature. To simplify calculations, we will restrict attention to the class of backgrounds whose twistor representatives h only have an $\bar{e}^0 \propto \mathrm{D}\bar\lambda$ component,

$$\mathsf{h} = \mathsf{h}_0\, \bar{e}^0 \in \Omega^{0,1}(\mathbb{PT}, \mathscr{O}(2))\,. \tag{8.2}$$

In particular, this covers the twistor representatives (7.24) of all SD radiative space-times. Integrability $\mathsf{T} = \bar\partial \mathsf{h} + \frac{1}{2}\{\mathsf{h}, \mathsf{h}\} = 0$ then reduces to holomorphicity in $\mu^{\dot\alpha}$,

$$\bar\partial_{\dot\alpha} \mathsf{h}_0 = 0 \implies \mathsf{h}_0 \equiv \mathsf{h}_0(\mu, \lambda, \hat\lambda)\,. \tag{8.3}$$

Note that h_0 is still allowed to have antiholomorphic dependence in λ.

Now perturb g_0 to $g = g_0 + \delta g$ such that g also satisfies the fully non-linear SD vacuum equations. Denote the corresponding space-time by $\mathcal{M}$ (of course, we have only deformed the metric; the underlying smooth manifold of $\mathcal{M}$ is the same as that of M). Its twistor representative is a linear perturbation $\mathsf{h} + h$ for some $\mathscr{O}(2)$-valued h satisfying (8.1). The $(1,0)$-forms of the deformed twistor complex structure read

$$\theta^{\dot\alpha} = \theta_0^{\dot\alpha} + \mathcal{L}_{\partial^{\dot\alpha}} h\,, \tag{8.4}$$

where $\theta_0^{\dot\alpha} = e^{\dot\alpha} + \mathcal{L}_{\partial^{\dot\alpha}} \mathsf{h}$ are now the background $(1,0)$-forms.

Let $y^{\alpha\dot\alpha}$ denote coordinates on M, which we continue to use over $\mathcal{M}$. Denote the background holomorphic curves by $\mathsf{Y} : \mu^{\dot\alpha} = \mathsf{F}^{\dot\alpha}(y, \lambda)$. Recall from (3.33) that they satisfy

$$\bar\partial_0 \mathsf{F}^{\dot\alpha} + \bar\partial_0 \lrcorner\, \mathcal{L}_{\partial^{\dot\alpha}} \mathsf{h}\big|_{\mathsf{Y}} = 0\,. \tag{8.5}$$

Turning on h deforms the background curves Y to new curves

$$Y: \quad \mu^{\dot\alpha} = F^{\dot\alpha}(y, \lambda) = \mathsf{F}^{\dot\alpha}(y, \lambda) + m^{\dot\alpha}(y, \lambda)\,. \tag{8.6}$$

Demanding $\bar\partial_0 \lrcorner\, (\theta^{\dot\alpha}|_Y) = 0$ shows that these satisfy a perturbed version of (3.44),

$$\bar\partial_0 F^{\dot\alpha} + \bar\partial_0 \lrcorner\, \mathcal{L}_{\partial^{\dot\alpha}} (\mathsf{h} + h)|_Y = 0 \tag{8.7}$$

Substituted for $F^{\dot\alpha}$ from (8.6) and $\bar\partial_0 \mathsf{F}^{\dot\alpha}$ from (8.5) gives us a PDE for the perturbation $m^{\dot\alpha}$,

$$\bar\partial_0 m^{\dot\alpha} + \bar\partial_0 \lrcorner \left(\mathcal{L}_{\partial^{\dot\alpha}} \mathsf{h}\big|_Y + \mathcal{L}_{\partial^{\dot\alpha}} h\big|_Y - \mathcal{L}_{\partial^{\dot\alpha}} \mathsf{h}\big|_{\mathsf{Y}}\right) = 0\,. \tag{8.8}$$

We can also make this more explicit by substituting $\mathcal{L}_{\partial_{\dot\alpha}} \mathsf{h} = \partial_{\dot\alpha} \mathsf{h}_0\, \bar{e}^0$.

Pick a spinor basis $\kappa_1^\alpha, \kappa_2^\alpha$ with normalization $\langle 1\, 2\rangle = -1$. Use it to define homogeneous coordinates on $\mathbb{P}^1$ and complex coordinates on M by

$$\lambda_\alpha = \lambda_1\, \kappa_{1\alpha} + \lambda_2\, \kappa_{2\alpha}\,, \qquad y^{\dot\alpha} := \mathsf{F}^{\dot\alpha}(y,\kappa_1)\,, \qquad \tilde y^{\dot\alpha} := \mathsf{F}^{\dot\alpha}(y,\kappa_2)\,. \tag{8.9}$$

Let D be the divisor $\{\lambda_1\lambda_2 = 0\} \subset \mathbb{PT}$. To solve (8.8), we will again impose the boundary conditions

$$m^{\dot\alpha}(y,\lambda) = \lambda_1\lambda_2\, \mathfrak{m}^{\dot\alpha}(y,\lambda) \in \Omega^{0,1}(\mathbb{PT}, \mathcal{O}(1)(-D)) \tag{8.10}$$

for some smooth $\mathcal{O}(-1)$-valued functions $\mathfrak{m}^{\dot\alpha}$. That is, $F^{\dot\alpha}(y,\kappa_1) = y^{\dot\alpha}$, $F^{\dot\alpha}(y,\kappa_2) = \tilde y^{\dot\alpha}$.

We can easily write a sigma model governing such holomorphic rational maps,

$$S_{\mathsf{h},h}[m](y) = \frac{1}{2\pi\mathrm{i}} \int_{\mathbb{P}^1} \frac{\mathrm{D}\lambda}{\lambda_1^2\lambda_2^2} \wedge \left(\frac{1}{2}\,[m\,\bar\partial m] + h|_Y + \mathsf{h}|_Y - \mathsf{h}|_{\mathsf{Y}} - \mathcal{L}_{\partial_{\dot\alpha}}\mathsf{h}\big|_{\mathsf{Y}}\, m^{\dot\alpha} \right), \tag{8.11}$$

where $\bar\partial \equiv \bar\partial|_{\mathbb{P}^1}$ as usual in such actions. As $\mathsf{h} \to 0$, this reduces to the original twistor sigma model action (3.49). A calculation analogous to the derivation of (3.55) shows that the equation of motion of $m^{\dot\alpha}$ reproduces (8.8) as long as h satisfies the boundary conditions

$$h \in \Omega^{0,1}(\mathbb{PT}, \mathcal{O}(2)(-2D))\,. \tag{8.12}$$

Meanwhile, the term $\mathsf{h}|_{\mathsf{Y}}$ seems superfluous as it does not depend on the field $m^{\dot\alpha}$. But a Taylor expansion around the background curves shows that

$$\begin{aligned} \mathsf{h}|_Y - \mathsf{h}|_{\mathsf{Y}} - \mathcal{L}_{\partial_{\dot\alpha}}\mathsf{h}\big|_{\mathsf{Y}}\, m^{\dot\alpha} &= (\mathsf{h}_0|_Y - \mathsf{h}_0|_{\mathsf{Y}} - \partial_{\dot\alpha}\mathsf{h}_0\big|_{\mathsf{Y}}\, m^{\dot\alpha})\,\bar e^0 \\ &= \sum_{p=2}^{\infty} \frac{1}{p!}\, m^{\dot\alpha_1}\cdots m^{\dot\alpha_p}\, \partial_{\dot\alpha_1}\cdots\partial_{\dot\alpha_p}\mathsf{h}_0|_{\mathsf{Y}}\,\bar e^0 \sim O(m^{\dot\alpha} m^{\dot\beta})\,, \end{aligned} \tag{8.13}$$

ensuring that the Lagrangian is non-singular at $\lambda = \kappa_1, \kappa_2$. Finally, just as in (3.61), the equation of motion of $\hat m^{\dot\alpha}$ imposes that the projection of $T = \overline{\nabla} h + \frac{1}{2}\{h,h\}$ along Y vanishes.

We will now relate the on-shell action of this background-coupled sigma model with the space-time metric $g = g_0 + \delta g$ on $\mathcal{M}$.

Proposition 8.1 *Let $h \in \Omega^{0,1}(\mathbb{PT}, \mathcal{O}(2)(-2D))$ satisfy $T = \overline{\nabla} h + \frac{1}{2}\{h,h\} = 0$ as above. Then it gives rise to an SD vacuum metric*

$$g = g_0 - 2\,\frac{\partial^2\Phi}{\partial y^{\dot\alpha}\partial\tilde y^{\dot\beta}}\, \mathrm{d}y^{\dot\alpha}\,\mathrm{d}\tilde y^{\dot\beta} \tag{8.14}$$

on the deformed space-time $\mathcal{M}$, with the scalar potential $\Phi(y)$ given by the on-shell action of our background-coupled twistor sigma model:

$$\Phi(y) = S_{\mathsf{h},h}[m](y)\big|_{\text{on-shell}}\,. \tag{8.15}$$

Proof The proof goes through the same steps as Theorem 3.2, so we will be brief. In what follows, let p^* and p^* denote pullbacks by the maps $\mathsf{p} : (y, \lambda) \mapsto Z^A = (\mathsf{F}^{\dot\alpha}, \lambda_\alpha)$ and $p : (y, \lambda) \mapsto Z^A = (F^{\dot\alpha}, \lambda_\alpha)$. A straightforward if tedious calculation shows that the $y^{\dot\alpha}$ derivative of the on-shell action reduces to

$$\frac{\partial}{\partial y^{\dot\alpha}} S_{\mathsf{h},h}[m](y)\big|_{\text{on-shell}} = \frac{1}{2\pi \mathrm{i}} \int_{\mathbb{P}^1} \frac{\mathrm{D}\lambda}{\lambda_1^2\lambda_2^2} \wedge \left(\frac{\partial m^{\dot\beta}}{\partial y^{\dot\alpha}} \bar\partial m_{\dot\beta} - \frac{\partial F^{\dot\beta}}{\partial y^{\dot\alpha}} \bar\partial m_{\dot\beta} - m_{\dot\beta} \bar\partial \frac{\partial \mathsf{F}^{\dot\beta}}{\partial y^{\dot\alpha}} \right), \tag{8.16}$$

having used the equation of motion (8.8) and the integrability conditions $T = 0$, and having dropped a total derivative term

$$\frac{1}{2\pi \mathrm{i}} \int_{\mathbb{P}^1} \frac{\mathrm{D}\lambda}{\lambda_1^2\lambda_2^2} \wedge \bar\partial \left(\frac{1}{2} \left[m \frac{\partial m}{\partial y^{\dot\alpha}} \right] + \frac{\partial}{\partial y^{\dot\alpha}} \lrcorner\, p^* h \right) \tag{8.17}$$

by virtue of the boundary conditions (8.10) and (8.12).

Expanding $F^{\dot\beta} = \mathsf{F}^{\dot\beta} + m^{\dot\beta}$ and $m^{\dot\beta} = \lambda_1\lambda_2\, \mathfrak{m}^{\dot\beta}$ shows that (8.16) is also a boundary term, but one that contributes non-trivially:

$$\begin{aligned} \frac{\partial}{\partial y^{\dot\alpha}} S_{\mathsf{h},h}[m](y)\big|_{\text{on-shell}} &= \frac{1}{2\pi \mathrm{i}} \int_{\mathbb{P}^1} \frac{\mathrm{D}\lambda}{\lambda_1^2\lambda_2^2} \wedge \bar\partial \left(m_{\dot\beta} \frac{\partial \mathsf{F}^{\dot\beta}}{\partial y^{\dot\alpha}} \right) \\ &= \frac{1}{2\pi \mathrm{i}} \int_{\mathbb{P}^1} \frac{\mathrm{D}\lambda}{\lambda_1\lambda_2} \wedge \bar\partial \left(\mathfrak{m}_{\dot\beta} \frac{\partial \mathsf{F}^{\dot\beta}}{\partial y^{\dot\alpha}} \right) \\ &= \mathfrak{m}_{\dot\beta} \frac{\partial \mathsf{F}^{\dot\beta}}{\partial y^{\dot\alpha}} \bigg|_{\lambda=\kappa_2} - \mathfrak{m}_{\dot\beta} \frac{\partial \mathsf{F}^{\dot\beta}}{\partial y^{\dot\alpha}} \bigg|_{\lambda=\kappa_1}. \end{aligned} \tag{8.18}$$

Recalling from (8.9) the definitions of $y^{\dot\alpha}$ and $\tilde y^{\dot\alpha}$, we see that only the second term survives. This leaves us with

$$\frac{\partial}{\partial y^{\dot\alpha}} S_{\mathsf{h},h}[m](y)\big|_{\text{on-shell}} = -\mathfrak{m}_{\dot\alpha}(y, \kappa_1) \tag{8.19}$$

Similarly, one can show that

$$\frac{\partial}{\partial \tilde y^{\dot\alpha}} S_{\mathsf{h},h}[m](y)\big|_{\text{on-shell}} = \mathfrak{m}_{\dot\alpha}(y, \kappa_2)\,, \tag{8.20}$$

just as we found for a flat background $\mathsf{h} = 0$ in (3.63).

Next, we substitute $m^{\dot\beta} = \lambda_1\lambda_2\, \mathfrak{m}^{\dot\beta}$ into (8.8) and use the Cauchy kernel of the Dolbeault operator $\bar\partial|_{\mathbb{P}^1} : \Omega^0(\mathbb{P}^1, \mathscr{O}(-1)) \to \Omega^{0,1}(\mathbb{P}^1, \mathscr{O}(-1))$ to find an integral equation

$$\mathfrak{m}^{\dot\alpha}(y, \lambda') = -\frac{1}{2\pi \mathrm{i}} \int_{\mathbb{P}^1} \frac{\mathrm{D}\lambda}{\langle \lambda'\, \lambda\rangle} \wedge \frac{\mathcal{L}_{\partial^{\dot\alpha}} h|_Y + \mathcal{L}_{\partial^{\dot\alpha}} \mathsf{h}|_Y - \mathcal{L}_{\partial^{\dot\alpha}} \mathsf{h}|_{\mathsf{Y}}}{\lambda_1\lambda_2}. \tag{8.21}$$

This is a perturbed version of (3.46). We can evaluate it at $\lambda' = \kappa_1$ to make (8.19) more explicit,

$$\frac{\partial}{\partial y^{\dot\alpha}} S_{\mathsf{h},h}[m](y)\Big|_{\text{on-shell}} = -\frac{1}{2\pi\mathrm{i}}\int_{\mathbb{P}^1}\frac{\mathrm{D}\lambda}{\lambda_1\,\lambda_2^2}\wedge\left(\mathcal{L}_{\partial^{\dot\alpha}}h|_Y + \mathcal{L}_{\partial^{\dot\alpha}}\mathsf{h}|_Y - \mathcal{L}_{\partial^{\dot\alpha}}\mathsf{h}|_{\mathsf{Y}}\right), \tag{8.22}$$

having remembered $\langle 1\,\lambda\rangle = -\lambda_2$. Staying steadfast on the course, we can compute its $\tilde y^{\dot\beta}$ derivative (or carefully recycle the result (3.68)) to find the simple final expression

$$\frac{\partial^2 S_{\mathsf{h},h}[m](y)}{\partial y^{\dot\alpha}\partial\tilde y^{\dot\beta}}\bigg|_{\text{on-shell}} = \frac{1}{2\pi\mathrm{i}}\int_Y\frac{\mathrm{D}\lambda\wedge\bar e^0}{\lambda_1\,\lambda_2^2}\left[\frac{\partial}{\partial\tilde y^{\dot\beta}}\lrcorner\,\bar\partial_0\lrcorner\, p^*\mathrm{d}\theta_{\dot\alpha} - \frac{\partial}{\partial\tilde y^{\dot\beta}}\lrcorner\,\bar\partial_0\lrcorner\,\mathsf{p}^*\mathrm{d}\theta_{0\dot\alpha}\right] \tag{8.23}$$

where the deformed and undeformed (1, 0)-forms $\theta^{\dot\alpha}$, $\theta_0^{\dot\alpha}$ are related by equation (8.4). In finding this, we have again dropped a boundary term containing h.

In Sect. 3.2, we proved that if the complex structure deformations are integrable, then the (1, 0)-forms on twistor space descend to the tetrad on space-time, up to an $\mathrm{SL}(2,\mathbb{C})$-valued spin frame defined by (3.39). In our case, we denote the background spin frame by $\mathsf{H}^{\dot\alpha}{}_{\dot\beta}$ and the perturbed one by $H^{\dot\alpha}{}_{\dot\beta}$. Then we find the undeformed and deformed versions of (3.39),

$$\begin{aligned}\mathsf{p}^*\theta_0^{\dot\alpha} &= \mathsf{H}^{\dot\alpha}{}_{\dot\beta}\,\lambda_\beta\,\boldsymbol{\theta}_0^{\beta\dot\beta} \qquad \operatorname{mod}\mathrm{D}\lambda\,,\\ p^*\theta^{\dot\alpha} &= H^{\dot\alpha}{}_{\dot\beta}\,\lambda_\beta\,\boldsymbol{\theta}^{\beta\dot\beta} \qquad \operatorname{mod}\mathrm{D}\lambda\,.\end{aligned} \tag{8.24}$$

Here, $\theta_0^{\dot\alpha} = e^{\dot\alpha} + \mathcal{L}_{\partial^{\dot\alpha}}\mathsf{h} = \lambda_\alpha\,\mathrm{d}x^{\alpha\dot\alpha} + \partial^{\dot\alpha}\mathsf{h}_0\,\bar e^0$. Since we demand $\bar\partial_0\lrcorner\,\mathsf{p}^*\theta_0^{\dot\alpha} = 0$ as the definition of p^*, the $\bar e^0$ components in $\mathsf{p}^*\theta_0^{\dot\alpha}$ get cancelled and we are left with only a space-time component in $\mathsf{p}^*\theta_0^{\dot\alpha}$,

$$\mathsf{p}^*\theta_0^{\dot\alpha} = \lambda_\alpha\,\mathrm{d}_y x^{\alpha\dot\alpha} = \mathrm{d}_y F^{\dot\alpha}(y,\lambda)\,. \tag{8.25}$$

Evaluating this at $\lambda = \kappa_1, \kappa_2$ and comparing with the first equation in (8.24) shows that

$$\boldsymbol{\theta}_0^{1\dot\beta} = -\mathsf{H}_{\dot\alpha}{}^{\dot\beta}(y,\kappa_1)\,\mathrm{d}y^{\dot\alpha}\,,\qquad \boldsymbol{\theta}_0^{2\dot\beta} = -\mathsf{H}_{\dot\alpha}{}^{\dot\beta}(y,\kappa_2)\,\mathrm{d}\tilde y^{\dot\alpha}\,, \tag{8.26}$$

having set $\boldsymbol{\theta}^{1\dot\alpha} = \kappa_{1\alpha}\boldsymbol{\theta}^{\alpha\dot\alpha}$, etc., and inverted the spin frames using $\epsilon_{\dot\alpha\dot\beta}\,\mathsf{H}^{\dot\alpha}{}_{\dot\gamma}\,\mathsf{H}^{\dot\beta}{}_{\dot\delta} = \epsilon_{\dot\gamma\dot\delta}$. An analogous analysis works for $\boldsymbol{\theta}^{\alpha\dot\alpha}$ owing to the fact that h vanishes at $\lambda = \kappa_1, \kappa_2$, leading to

$$\boldsymbol{\theta}^{1\dot\beta} = -H_{\dot\alpha}{}^{\dot\beta}(y,\kappa_1)\,\mathrm{d}y^{\dot\alpha}\,,\qquad \boldsymbol{\theta}^{2\dot\beta} = -H_{\dot\alpha}{}^{\dot\beta}(y,\kappa_2)\,\mathrm{d}\tilde y^{\dot\alpha}\,. \tag{8.27}$$

Having determined the tetrad, the undeformed and deformed space-time metrics read

$$g_0 = 2\,\mathsf{H}_{\dot\alpha\dot\gamma}(y,\kappa_1)\,\mathsf{H}_{\dot\beta}{}^{\dot\gamma}(y,\kappa_2)\,\mathrm{d}y^{\dot\alpha}\,\mathrm{d}\tilde y^{\dot\beta}\,,\qquad g = 2\,H_{\dot\alpha\dot\gamma}(y,\kappa_1)\,H_{\dot\beta}{}^{\dot\gamma}(y,\kappa_2)\,\mathrm{d}y^{\dot\alpha}\,\mathrm{d}\tilde y^{\dot\beta}\,. \tag{8.28}$$

The last step is to use fix the spin frames in terms of the twistor sigma model action.

Substituting (8.24) into (8.23), we can simply recycle the computation in (3.73), once for M and then for $\mathcal{M}$. This yields

$$\begin{aligned}\left.\frac{\partial^2 S_{\mathsf{h},h}[m](y)}{\partial y^{\dot\alpha}\partial\tilde y^{\dot\beta}}\right|_{\text{on-shell}} &= \left(\epsilon_{\dot\alpha\dot\beta} + H_{\dot\alpha}{}^{\dot\gamma}(y,\kappa_1)\,H_{\dot\beta\dot\gamma}(y,\kappa_2)\right) - \left(\epsilon_{\dot\alpha\dot\beta} + \mathsf{H}_{\dot\alpha}{}^{\dot\gamma}(y,\kappa_1)\,\mathsf{H}_{\dot\beta\dot\gamma}(y,\kappa_2)\right)\\ &= H_{\dot\alpha}{}^{\dot\gamma}(y,\kappa_1)\,H_{\dot\beta\dot\gamma}(y,\kappa_2) - \mathsf{H}_{\dot\alpha}{}^{\dot\gamma}(y,\kappa_1)\,\mathsf{H}_{\dot\beta\dot\gamma}(y,\kappa_2)\,.\end{aligned} \tag{8.29}$$

Using this, Eq. (8.28) determines the deformed metric in terms of the metric on M,

$$\begin{aligned} g &= 2\left(\mathsf{H}_{\dot\alpha\dot\gamma}(y,\kappa_1)\,\mathsf{H}_{\dot\beta}{}^{\dot\gamma}(y,\kappa_2) - \left.\frac{\partial^2 S_{\mathsf{h},h}[m](y)}{\partial y^{\dot\alpha}\partial\tilde y^{\dot\beta}}\right|_{\text{on-shell}}\right)\mathrm{d}y^{\dot\alpha}\,\mathrm{d}\tilde y^{\dot\beta}\\ &= g_0 - \frac{\partial^2\Phi}{\partial y^{\dot\alpha}\partial\tilde y^{\dot\beta}}\,\mathrm{d}y^{\dot\alpha}\,\mathrm{d}\tilde y^{\dot\beta}\,,\end{aligned} \tag{8.30}$$

having expressed this in terms of a scalar potential $\Phi(y) = S_{\mathsf{h},h}[m](y)|_{\text{on-shell}}$. If the metric g_0 can be written as a Kähler metric in the coordinates $y^{\dot\alpha}$, $\tilde y^{\dot\alpha}$, then this scalar potential can be thought of as the corresponding Kähler scalar perturbation. □

To compute amplitudes, we will also need expressions for the ASD 2-forms. Just like the metric, the deformed 2-form $\Sigma^{12} = \boldsymbol{\theta}^{1\dot\alpha}\wedge\boldsymbol{\theta}^2{}_{\dot\alpha}$ can be related to the background 2-form $\Sigma_0^{12} = \boldsymbol{\theta}_0^{1\dot\alpha}\wedge\boldsymbol{\theta}_0{}^2{}_{\dot\alpha}$ by applying (8.26), (8.27) and (8.29):

$$\begin{aligned}\Sigma^{12} &= H_{\dot\alpha}{}^{\dot\gamma}(y,\kappa_1)\,H_{\dot\beta\dot\gamma}(y,\kappa_2)\,\mathrm{d}y^{\dot\alpha}\wedge\mathrm{d}\tilde y^{\dot\beta}\\ &= \Sigma_0^{12} - \left.\frac{\partial^2 S_{\mathsf{h},h}[m](y)}{\partial y^{\dot\alpha}\partial\tilde y^{\dot\beta}}\right|_{\text{on-shell}}\mathrm{d}y^{\dot\alpha}\wedge\mathrm{d}\tilde y^{\dot\beta}\,.\end{aligned} \tag{8.31}$$

This will help us relate the generating functional for MHV amplitudes on SD radiative space-times to semi-classical correlators of the background-coupled sigma models.

8.2 MHV Amplitudes

In Eq. (5.9), we described a generating functional that computes tree-level MHV graviton amplitudes around SD backgrounds M [1]. We repeat this here for convenience:

$$\mathcal{G}(1,2) \equiv \int_{\mathcal{M}} \Sigma^{\alpha\beta}\wedge\gamma_{1\,\alpha}{}^{\delta}\wedge\gamma_{2\,\delta\beta}\,. \tag{8.32}$$

Here, the positive helicity gravitons have been encoded into the geometry of a deformed SD vacuum space-time $\mathcal{M} = M\oplus\{+\text{ helicity gravitons}\}$. Their information is contained in the deformed ASD 2-forms $\Sigma^{\alpha\beta}$. On the other hand, the negative helicity gravitons $1, 2$ are separately captured by the factors γ_1, γ_2 denoting spin connection perturbations. Let us now put this to use to finally compute some new amplitudes!

Lift to twistor space. As discussed in Sect. 7.3, SD and ASD perturbations to the SD radiative background $\mathcal{M}$ can be represented as momentum eigenstates with momenta $k_i^{\alpha\dot{\alpha}} = \kappa_i^\alpha \tilde{\kappa}_i^{\dot{\alpha}}$. The ASD spin connection perturbations can be taken to be

$$\gamma_i^{\alpha\beta}(y) = -\frac{2\,\mathrm{i}\,\tilde{\xi}^{\dot{\alpha}}}{[\tilde{\xi}\,i]}\,\kappa_i^\alpha\,\kappa_i^\beta\,\mathrm{e}^{\mathrm{i}\phi_i}\,\mathrm{d}_y\mathsf{F}_{\dot{\alpha}}(y,\kappa_i)\,, \qquad \phi_i(y) \equiv [\mathsf{F}(y,\kappa_i)\,i]\,, \tag{8.33}$$

for $i = 1, 2$. The constant spinor $\tilde{\xi}_{\dot{\alpha}}$ encodes residual gauge (diffeomorphism) freedom in the representation of the spin connection perturbations and drop out of invariant quantities like Weyl curvature perturbations. Applying (7.30), the wavefunctions (8.33) are seen to satisfy the required linearized field equations $\mathrm{d}\gamma_i^{\alpha\beta} = \psi_i^{\alpha\beta\gamma\delta}\,\Sigma_{0\gamma\delta}$:

$$\begin{aligned} \mathrm{d}\gamma_i^{\alpha\beta} &= -\frac{2\,\tilde{\kappa}_i^{\dot{\alpha}}\,\tilde{\xi}^{\dot{\beta}}}{[\tilde{\xi}\,i]}\,\kappa_i^\alpha\,\kappa_i^\beta\,\mathrm{e}^{\mathrm{i}\phi_i}\,\mathrm{d}_y\mathsf{F}_{\dot{\alpha}}(y,\kappa_i)\wedge\mathrm{d}_y\mathsf{F}_{\dot{\beta}}(y,\kappa_i) \\ &= \kappa_i^\alpha\,\kappa_i^\beta\,\mathrm{e}^{\mathrm{i}\phi_i}\,\mathrm{d}_y\mathsf{F}^{\dot{\alpha}}(y,\kappa_i)\wedge\mathrm{d}_y\mathsf{F}_{\dot{\alpha}}(y,\kappa_i) = \kappa_i^\alpha\,\kappa_i^\beta\,\kappa_i^\gamma\,\kappa_i^\delta\,\mathrm{e}^{\mathrm{i}\phi_i}\,\Sigma_{0\gamma\delta} \end{aligned} \tag{8.34}$$

with curvatures of the form (7.53) as predicted by the Penrose transform.

The expressions for the ASD perturbations are simplified by using the complex coordinates $(y^{\dot{\alpha}}, \tilde{y}^{\dot{\alpha}})$ of (8.9); these are adapted to the ASD perturbations by identifying the spinor basis κ_1, κ_2 with the undotted momentum spinors of the negative helicity gravitons. In these coordinates (8.33) become

$$\gamma_1^{\alpha\beta} = -2\,\mathrm{i}\,\frac{[\tilde{\xi}\,\mathrm{d}y]}{[\tilde{\xi}\,1]}\,\kappa_1^\alpha\,\kappa_1^\beta\,\mathrm{e}^{\mathrm{i}[y\,1]}\,, \qquad \gamma_2^{\alpha\beta} = -2\,\mathrm{i}\,\frac{[\tilde{\xi}\,\mathrm{d}\tilde{y}]}{[\tilde{\xi}\,2]}\,\kappa_2^\alpha\,\kappa_2^\beta\,\mathrm{e}^{\mathrm{i}[\tilde{y}\,2]}\,, \tag{8.35}$$

which take on a form that closely resembles negative helicity gravitons in flat space.

Just as in our flat space calculation (5.20), inserting these into the MHV generating functional (8.32) and using (8.31) leads to

$$\mathcal{G}(1,2) = \mathcal{G}_{\text{2-pt}}(1,2) + \int_{\mathcal{M}} \mathrm{d}^2y\,\mathrm{d}^2\tilde{y}\;\mathrm{e}^{\mathrm{i}[y\,1]+\mathrm{i}[\tilde{y}\,2]}\;S_{\mathsf{h},h}[m](y)\Big|_{\text{on-shell}}\,, \tag{8.36}$$

Here, the first term on the right computes a 2-point amplitude of negative helicity graviton states on the SD background M,

$$\mathcal{G}_{\text{2-pt}}(1,2) = \int_M \Sigma_0^{\alpha\beta}\wedge\gamma_{1\,\alpha}{}^{\delta}\wedge\gamma_{2\,\delta\beta}\,. \tag{8.37}$$

On self-dual backgrounds, the linearized field Eqs. (5.6), (5.7) and the linearization $\Sigma_0^{(\alpha\beta} \wedge \sigma_i^{\gamma\delta)} = 0, i = 1, 2,$ of the simplicity constraint $\Sigma^{(\alpha\beta} \wedge \Sigma^{\gamma\delta)} = 0$ can be used to reduce this to a boundary term,

$$\mathcal{G}_{2\text{-pt}}(1,2) = \frac{1}{2}\int_M \mathrm{d}\left(\sigma_1^{\alpha\beta} \wedge \gamma_{2\,\alpha\beta}\right) = 0\,, \tag{8.38}$$

which vanishes by Stokes' theorem (see [1] for a more careful boundary analysis in Lorentzian signature). This is a feature reflecting the integrability underlying self-dual backgrounds. Thus, the MHV generating functional and its perturbative expansion are entirely controlled by the on-shell action of the twistor sigma model.

MHV amplitudes on SD radiative backgrounds. Armed with the generating functional (8.36), we are now in a position to perturbatively expand the SD space-time $\mathcal{M}$ around the background M. In particular, to recover the n-point tree-level MHV amplitude on the SD radiative background M, $\mathcal{G}(1,2)$ must be expanded to order $n-2$ in h. This corresponds to extracting the $(n-2)$-linear piece of the on-shell twistor sigma model action $S_{\mathsf{h},h}$. Since on-shell actions are computed by tree-level Feynman graphs, this multi-linear contribution can be expressed in terms of a connected, tree-level correlation function in the QFT on $\mathbb{P}^1$ defined by the twistor sigma model.

To bring the background-coupled sigma model (8.11) to a form amenable to perturbation theory around M, we first Taylor expand $\mathsf{h}|_Y$ around the background curves Y as in (8.13),

$$\begin{aligned} S_{\mathsf{h},h}[m](y) = \frac{1}{2\pi \mathrm{i}}\int_{\mathbb{P}^1} \frac{\mathrm{D}\lambda}{\langle\lambda\, 1\rangle^2\,\langle\lambda\, 2\rangle^2} \wedge \bigg(\frac{1}{2}\,[m\,\bar\partial m] + h|_Y \\ + \sum_{p=2}^{\infty}\frac{1}{p!}\, m^{\dot\alpha_1}\cdots m^{\dot\alpha_p}\,\partial_{\dot\alpha_1}\cdots\partial_{\dot\alpha_p}\mathsf{h}_0|_{\mathsf{Y}}\,\bar e^0\bigg)\,, \end{aligned} \tag{8.39}$$

having recalled $\lambda_1 = -\langle\lambda\, 2\rangle$, $\lambda_2 = \langle\lambda\, 1\rangle$. Letting $h = \sum_{i=3}^n a_i\, h_i$, the MHV amplitude is defined as in Eq. (5.10). The piece of the on-shell action that is multi-linear in each h_i is extracted by the a_i derivatives

$$\begin{aligned} \frac{1}{2\pi \mathrm{i}}\left(\prod_i \frac{\partial}{\partial a_i}\right)\int_{\mathbb{P}^1} \frac{\mathrm{D}\lambda}{\langle\lambda\, 1\rangle^2\,\langle\lambda\, 2\rangle^2} \wedge \bigg(\frac{1}{2}\,[m\,\bar\partial m] + \sum_{i=3}^{n} a_i\, h_i|_Y \\ + \sum_{p=2}^{\infty}\frac{1}{p!}\, m^{\dot\alpha_1}\cdots m^{\dot\alpha_p}\,\partial_{\dot\alpha_1}\cdots\partial_{\dot\alpha_p}\mathsf{h}_0|_{\mathsf{Y}}\,\bar e^0\bigg)\bigg|_{a_i=0} = \left\langle \prod_{i=3}^{n} V_i\right\rangle^0_{\text{tree}}\,, \end{aligned} \tag{8.40}$$

where the expectation value in the final line denotes a sum of connected tree diagrams (i.e., diagrams of $\mathcal{O}(\hbar^0)$) in the "background" $\mathbb{P}^1$ theory

$$S_{\mathsf{h},0}[m](y) = \frac{1}{2\pi i\hbar}\int_{\mathbb{P}^1}\frac{\mathrm{D}\lambda\wedge\bar{e}^0}{\langle\lambda\,1\rangle^2\,\langle\lambda\,2\rangle^2}\left(\frac{1}{2}\,[m\,\bar{\eth}_0 m] + \sum_{p=2}^{\infty}\frac{m^{\dot\alpha_1}\cdots m^{\dot\alpha_p}}{p!}\,\partial_{\dot\alpha_1}\cdots\partial_{\dot\alpha_p}\mathsf{h}_0|_{\mathsf{Y}}\right), \tag{8.41}$$

with vertex operators

$$V_i := \frac{1}{2\pi i\hbar}\int_{\mathbb{P}^1}\frac{\mathrm{D}\lambda\wedge h_i|_Y}{\langle\lambda\,1\rangle^2\,\langle\lambda\,2\rangle^2} \tag{8.42}$$

encoding the positive helicity gravitons. It will be assumed that each individual graviton is represented by a momentum eigenstate:

$$h_i(Z) = \int_{\mathbb{C}^*}\frac{\mathrm{d}s}{s^3}\,\bar\delta^2(\kappa_i - s\,\lambda)\,\mathrm{e}^{\mathrm{i}s[\mu\,i]}, \tag{8.43}$$

with on-shell momentum $k_i^{\alpha\dot\alpha} = \kappa_i^\alpha\tilde\kappa_i^{\dot\alpha}$. Evaluating V_i on such a state results in the momentum eigenstate vertex operator

$$V_i = \frac{1}{2\pi i\hbar}\,\frac{\mathrm{e}^{\mathrm{i}[\mathsf{F}(y,\kappa_i)\,i]}\,\mathrm{e}^{\mathrm{i}[m(y,\kappa_i)\,i]}}{\langle 1\,i\rangle^2\,\langle 2\,i\rangle^2} = \frac{1}{2\pi i\hbar}\,\frac{\mathrm{e}^{\mathrm{i}\phi_i(y)}\,\mathrm{e}^{\mathrm{i}[m(y,\kappa_i)\,i]}}{\langle 1\,i\rangle^2\,\langle 2\,i\rangle^2}, \tag{8.44}$$

having pulled back to the curve $Y:\mu^{\dot\alpha} = \mathsf{F}^{\dot\alpha} + m^{\dot\alpha}$ and performed all integrals against the delta functions in the h_i. Also, we have abbreviated $[\mathsf{F}(y,\kappa_i)\,i] \equiv \phi_i(y)$ as in previous sections.

The $h = 0$ sigma model action (8.41) depends only on the fixed SD radiative background M through h. However, there are infinitely many terms of higher and higher valence in $m^{\dot\alpha}$ appearing in this action, as curves Y corresponding to points $y \in \mathcal{M}$ are expanded around curves Y corresponding to points $y \in M$. The quadratic term in this expansion simply defines the kinetic term of the action, while the contribution of all higher-order terms to the correlator (8.40) can be accounted for by absorbing them into *background* vertex operators:

$$U^{(p)} := \frac{1}{2\pi i\hbar}\,\frac{1}{p!}\int_{\mathbb{P}^1}\frac{\mathrm{D}\lambda\wedge\bar{e}^0}{\langle\lambda\,1\rangle^2\,\langle\lambda\,2\rangle^2}\,\partial_{\dot\alpha_1}\cdots\partial_{\dot\alpha_p}\mathsf{h}_0\big|_{\mathsf{Y}}\,m^{\dot\alpha_1}\cdots m^{\dot\alpha_p}, \qquad p\geq 3. \tag{8.45}$$

Taking into account the relationship (7.24) between h and the characteristic radiative data of M, these background vertex operators can be written equivalently as:

$$U^{(p)} = \frac{1}{2\pi i\hbar}\,\frac{1}{p!}\int_{\mathbb{P}^1}\frac{\mathrm{D}\lambda\wedge\mathrm{D}\bar\lambda}{\langle\lambda\,1\rangle^2\,\langle\lambda\,2\rangle^2}\,N^{(p-2)}(u,\lambda,\bar\lambda)\,[m\,\bar\lambda]^p, \tag{8.46}$$

where $u = [\mu\,\bar\lambda] = [\mathsf{F}\,\bar\lambda]$ and $N^{(k)} := \partial_u^k N = -\partial_u^{k+1}\tilde{\boldsymbol{\sigma}}^0$ is the kth u-derivative of the news function of M.

Since these background vertex operators do not introduce any new powers of the h_i, arbitrarily many of them can be brought down into the correlation function (8.40).

Thus, the problem of computing the MHV amplitudes of $n-2$ positive helicity gravitons on M is reduced to computing the connected tree-level correlation function

$$\sum_{t=0}^{\infty}\sum_{p_1,\ldots,p_t}\left\langle\prod_{i=3}^{n}V_i\prod_{\mathrm{m}=1}^{t}U^{(p_{\mathrm{m}})}\right\rangle^0_{\text{tree}}, \tag{8.47}$$

in the free QFT on $\mathbb{P}^1$ defined by

$$S[m]=\frac{1}{4\pi\mathrm{i}\hbar}\int_{\mathbb{P}^1}\frac{\mathrm{D}\lambda\wedge\bar{e}^0}{\langle\lambda\,1\rangle^2\,\langle\lambda\,2\rangle^2}\left[m^{\dot\alpha}\left(\epsilon_{\dot\beta\dot\alpha}\bar\partial_0+\partial_{\dot\alpha}\partial_{\dot\beta}\mathsf{h}_0\right)m^{\dot\beta}\right]. \tag{8.48}$$

In (8.47), the second sum is over all $p_1,\ldots,p_t\geq 3$; that is, over all possible 'valences' of the background vertex operators. Physically, the role of these background vertex operators is clear: the $t>0$ terms in (8.47) correspond to tail terms in the MHV amplitude which arise because of the failure of Huygens' principle for graviton scattering in any curved background [2, 3].

The computation of (8.47) now proceeds making use of the OPE of the $m^{\dot\alpha}$ fields defined by the quadratic action (8.48):

$$m^{\dot\alpha}(\lambda)\,m^{\dot\beta}(\lambda')\sim 2\pi\mathrm{i}\hbar\,\frac{\mathsf{H}^{\dot\alpha}{}_{\dot\gamma}(y,\lambda)\,\mathsf{H}^{\dot\beta\dot\gamma}(y,\lambda')}{\langle\lambda\,\lambda'\rangle}\,\langle\lambda\,1\rangle\,\langle\lambda\,2\rangle\,\langle\lambda'\,1\rangle\,\langle\lambda'\,2\rangle\,. \tag{8.49}$$

This is obtained from dressing the OPE of the $\mathsf{h}=0$ free theory with factors of dotted spin frames. Specializing the PDE (3.41) to representatives $\mathsf{h}=\mathsf{h}_0\,\bar{e}^0$, we find that $\mathsf{H}^{\dot\alpha}{}_{\dot\beta}$ satisfies

$$\bar\partial_0\mathsf{H}^{\dot\alpha}{}_{\dot\beta}+\partial^{\dot\alpha}\partial_{\dot\gamma}\mathsf{h}_0\,\mathsf{H}^{\dot\gamma}{}_{\dot\beta}=0\,. \tag{8.50}$$

As a result, the dressed OPE (8.49) satisfies the equation of motion that follows from varying (8.48). The right-hand-side of this OPE acts as a Green's function for the Dolbeault operator on the dotted spinor bundles over background curves Y, and the appearance of the momentum spinors κ_1^α, κ_2^α in the OPE represents a gauge choice for the inverse of the $\bar\partial$-operator acting on sections of $\mathcal{O}(1)\to\mathbb{P}^1$. While $m^{\dot\alpha}$ is valued in $\mathcal{O}(1)$, the boundary conditions (8.10) ensure that it has no zero modes, so all insertions of $m^{\dot\alpha}$ in the correlator must be Wick contracted away using (8.49) to obtain non-vanishing contributions.

As a warm-up, it is illustrative to first consider the $t=0$ term in (8.47), which contains only external graviton vertex operators. This is given by

$$\left\langle\prod_{j=3}^{n}V_j\right\rangle^0_{\text{tree}}=\left\langle\prod_{j=3}^{n}\mathrm{e}^{\mathrm{i}[m(y,\kappa_j)\,j]}\right\rangle^0_{\text{tree}}\prod_{l=3}^{n}\frac{\mathrm{e}^{\mathrm{i}\phi_l(y)}}{\langle 1\,l\rangle^2\,\langle 2\,l\rangle^2}, \tag{8.51}$$

having inserted the explicit vertex operators (8.44). In this case, the computation boils down to summing all connected tree graphs on $n-2$ vertices, with the weights assigned to each edge given by the OPE (8.49) acting between two vertex operators, say i and j:

$$\begin{aligned}&\frac{[\![i\,j]\!]}{\langle i\,j\rangle}\,\langle i\,1\rangle\,\langle i\,2\rangle\,\langle j\,1\rangle\,\langle j\,2\rangle\,,\\ &\text{where}\quad [\![i\,j]\!](y) := [\tilde{K}_i\,\tilde{K}_j](y) = \tilde{\kappa}_{i\dot{\alpha}}\,\mathsf{H}^{\dot{\alpha}\dot{\gamma}}(y,\kappa_i)\,\tilde{\kappa}_{j\dot{\beta}}\,\mathsf{H}^{\dot{\beta}}{}_{\dot{\gamma}}(y,\kappa_j)\,.\end{aligned} \tag{8.52}$$

This arises from (8.49) on computing the tree OPE of two exponentials $\mathrm{e}^{\mathrm{i}[m(y,\kappa_i)\,i]}\,\mathrm{e}^{\mathrm{i}[m(y,\kappa_j)\,j]}$, i.e., on keeping the term with just a single Wick contraction. As in (8.49), each such OPE also comes with a factor of $2\pi\mathrm{i}\hbar$. All such factors cancel against the $1/2\pi\mathrm{i}\hbar$ prefactors in the vertex operators precisely for tree diagrams, so that loop diagrams drop out in the classical $\hbar \to 0$ limit as required. So we will drop such factors in all our subsequent expressions.

As with the flat space calculation in Lemma 5.2, the sum of these tree-level Feynman diagrams is accomplished using the weighted matrix-tree theorem (cf., [4–6]):

$$\left\langle \prod_{j=3}^{n} V_j \right\rangle^0_{\text{tree}} = |\mathcal{L}^i_i| \prod_{j=3}^{n} \frac{\mathrm{e}^{\mathrm{i}\phi_j(y)}}{\langle 1\,j\rangle^2\,\langle 2\,j\rangle^2}\,, \tag{8.53}$$

where $\mathcal{L}$ is the weighted Laplacian matrix whose off-diagonal entries are given by (8.52) and diagonal entries are

$$\mathcal{L}_{ii} = -\sum_{j\neq i} \frac{[\![i\,j]\!]}{\langle i\,j\rangle}\,\langle i\,1\rangle\,\langle i\,2\rangle\,\langle j\,1\rangle\,\langle j\,2\rangle\,, \tag{8.54}$$

and $|\mathcal{L}^i_i|$ denotes the minor of $\mathcal{L}$ obtained by removing the row and column corresponding to some external graviton i. The weighted matrix-tree theorem ensures that the equality (8.53) is independent of the choice of $i \in \{3, \ldots, n\}$.

To make the connection to known formulae for graviton scattering in vacuum transparent, it is useful to rewrite (8.53) by taking out a factor of $\langle \lambda_i\,1\rangle\,\langle \lambda_i\,2\rangle$ from each row and column of the determinant to give

$$\left\langle \prod_{j=3}^{n} V_j \right\rangle^0_{\text{tree}} = \frac{|\mathbb{H}^i_i|}{\langle 1\,i\rangle^2\,\langle 2\,i\rangle^2} \prod_{j=3}^{n} \mathrm{e}^{\mathrm{i}\phi_j(y)}\,, \tag{8.55}$$

where $\mathbb{H}$ is the $(n-2)\times(n-2)$ matrix with entries

$$\mathbb{H}_{ij} = \frac{[\![i\,j]\!]}{\langle i\,j\rangle}\,,\quad i \neq j\,,\qquad \mathbb{H}_{ii} = -\sum_{j\neq i} \frac{[\![i\,j]\!]}{\langle i\,j\rangle}\,\frac{\langle 1\,j\rangle\,\langle 2\,j\rangle}{\langle 1\,i\rangle\,\langle 2\,i\rangle}\,, \tag{8.56}$$

which is a background-dressed analogue of the $n \times n$ matrix (5.42) appearing in Hodges' formula for MHV graviton scattering in flat space.

Now consider a generic term with fixed $t > 0$ and $p_1, \ldots, p_t \geq 3$ in (8.47). Each insertion of a background vertex operator $U^{(p_{\mathrm{m}})}$ comes with at least 3 insertions of $m^{\dot{\alpha}}$, all of which must be Wick contracted for the correlator to be non-vanishing. These $m^{\dot{\alpha}}$-insertions can contract against the other background vertex operators or the $n-2$ positive helicity graviton vertex operators. It is easy to see that such Wick contractions cannot saturate the $m^{\dot{\alpha}}$-insertions via tree diagrams unless $t \leq (n-2) - 2 = n - 4$, providing an upper bound on the number of background vertex insertions.

Returning to the full MHV amplitude (8.47), to compute a correlator like

$$\left\langle \prod_{i=3}^{n} V_i \prod_{\mathrm{m}=1}^{t} U^{(p_{\mathrm{m}})} \right\rangle_{\text{tree}}^{0} \tag{8.57}$$

we must now sum over all connected tree diagrams on $n - 2 + t$ vertices where the valence of the t background vertices is fixed to be $p_1, \ldots, p_{\mathrm{m}}$. This can be reduced to summing over tree diagrams without any restriction on the valence by the following trick. We can represent the background vertex operator (8.46) as a "soft mode" of a momentum eigenstate,

$$U^{(p)} = \frac{1}{2\pi \mathrm{i} \hbar} \int_{\mathbb{P}^1} \frac{\mathrm{D}\lambda \wedge \mathrm{D}\bar{\lambda}}{\langle \lambda\, 1\rangle^2 \langle \lambda\, 2\rangle^2} N^{(p-2)}(u, \lambda, \bar{\lambda}) \frac{\mathrm{i}^{-p}}{p!} \frac{\partial^p}{\partial \varepsilon^p} \mathrm{e}^{\mathrm{i}\varepsilon[m\, \bar{\lambda}]} \Big|_{\varepsilon=0}. \tag{8.58}$$

The dummy variable ε acts as a fictitious "energy" of the background vertex, and the valence p now becomes the "order" of a "soft limit" $\varepsilon \to 0$. With the representation (8.58) of the background vertices, the tree correlator (8.57) reduces to

$$\begin{aligned} &\prod_{j=3}^{n} \frac{\mathrm{e}^{\mathrm{i}\phi_j}}{\langle 1\, j\rangle^2 \langle 2\, j\rangle^2} \prod_{\mathrm{m}=1}^{t} \int_{\mathbb{P}^1} \frac{\mathrm{D}\lambda_{\mathrm{m}} \wedge \mathrm{D}\bar{\lambda}_{\mathrm{m}}}{\langle 1\, \lambda_{\mathrm{m}}\rangle^2 \langle 2\, \lambda_{\mathrm{m}}\rangle^2} N_{\mathrm{m}}^{(p_{\mathrm{m}}-2)} \frac{\mathrm{i}^{-p_{\mathrm{m}}}}{p_{\mathrm{m}}!} \frac{\partial^{p_{\mathrm{m}}}}{\partial \varepsilon_{\mathrm{m}}^{p_{\mathrm{m}}}} \\ &\times \left\langle \prod_{j=3}^{n} \mathrm{e}^{\mathrm{i}[m(y,\kappa_j)\, j]} \prod_{\mathrm{m}=1}^{t} \mathrm{e}^{\mathrm{i}\varepsilon_{\mathrm{m}}[m(y,\lambda_{\mathrm{m}})\, \bar{\lambda}_{\mathrm{m}}]} \right\rangle_{\text{tree}}^{0} \Bigg|_{\varepsilon_{\mathrm{m}}=0\, \forall\, \mathrm{m}}, \end{aligned} \tag{8.59}$$

where $N_{\mathrm{m}} := N([\mathsf{F}(y, \lambda_{\mathrm{m}})\, \bar{\lambda}_{\mathrm{m}}], \lambda_{\mathrm{m}}, \bar{\lambda}_{\mathrm{m}})$ denotes copies of the news function of the SD radiative background M pulled back to the twistor curve. This can now be evaluated in the same manner as (8.51).

Recycling the result (8.55) by including such "soft graviton" tails yields

$$\begin{aligned} &\left\langle \prod_{j=3}^{n} \mathrm{e}^{\mathrm{i}[m(y,\kappa_j)\, j]} \prod_{\mathrm{m}=1}^{t} \mathrm{e}^{\mathrm{i}\varepsilon_{\mathrm{m}}[m(y,\lambda_{\mathrm{m}})\, \bar{\lambda}_{\mathrm{m}}]} \right\rangle_{\text{tree}}^{0} \Bigg|_{\varepsilon_{\mathrm{m}}=0} \\ &= \frac{|\mathcal{H}^i_i|}{\langle 1\, i\rangle^2 \langle 2\, i\rangle^2} \prod_{j=3}^{n} \langle 1\, j\rangle^2 \langle 2\, j\rangle^2 \prod_{\mathrm{m}=1}^{t} \langle 1\, \lambda_{\mathrm{m}}\rangle^2 \langle 2\, \lambda_{\mathrm{m}}\rangle^2. \end{aligned} \tag{8.60}$$

Here, we have introduced the $(n+t-2) \times (n+t-2)$ matrix $\mathcal{H}$ with block decomposition

$$\mathcal{H} = \begin{pmatrix} \mathbb{H} & \mathfrak{h} \\ \mathfrak{h}^{\mathrm{T}} & \mathbb{T} \end{pmatrix}, \tag{8.61}$$

and entries

$$\mathbb{H}_{ij} = \frac{[\![i\,j]\!]}{\langle i\,j\rangle}, \quad i \neq j \tag{8.62}$$

$$\mathbb{H}_{ii} = -\sum_{j\neq i} \frac{[\![i\,j]\!]}{\langle i\,j\rangle} \frac{\langle 1\,j\rangle\,\langle 2\,j\rangle}{\langle 1\,i\rangle\,\langle 2\,i\rangle} - \sum_{\mathrm{m}=1}^{t} \varepsilon_{\mathrm{m}} \frac{[\![i\,\bar{\lambda}_{\mathrm{m}}]\!]}{\langle i\,\lambda_{\mathrm{m}}\rangle} \frac{\langle 1\,\lambda_{\mathrm{m}}\rangle\,\langle 2\,\lambda_{\mathrm{m}}\rangle}{\langle 1\,i\rangle\,\langle 2\,i\rangle},$$

$$\mathfrak{h}_{i\mathrm{m}} = \varepsilon_{\mathrm{m}} \frac{[\![i\,\bar{\lambda}_{\mathrm{m}}]\!]}{\langle i\,\lambda_{\mathrm{m}}\rangle}, \qquad \mathbb{T}_{\mathrm{mn}} = \varepsilon_{\mathrm{m}}\varepsilon_{\mathrm{n}} \frac{[\![\bar{\lambda}_{\mathrm{m}}\,\bar{\lambda}_{\mathrm{n}}]\!]}{\langle \lambda_{\mathrm{m}}\,\lambda_{\mathrm{n}}\rangle}, \quad \mathrm{m} \neq \mathrm{n},$$

$$\mathbb{T}_{\mathrm{mm}} = -\varepsilon_{\mathrm{m}} \sum_{i=1}^{n} \frac{[\![\bar{\lambda}_{\mathrm{m}}\,i]\!]}{\langle \lambda_{\mathrm{m}}\,i\rangle} \frac{\langle 1\,i\rangle\,\langle 2\,i\rangle}{\langle 1\,\lambda_{\mathrm{m}}\rangle\,\langle 2\,\lambda_{\mathrm{m}}\rangle} - \varepsilon_{\mathrm{m}} \sum_{\mathrm{n}\neq\mathrm{m}} \varepsilon_{\mathrm{n}} \frac{[\![\bar{\lambda}_{\mathrm{m}}\,\bar{\lambda}_{\mathrm{n}}]\!]}{\langle \lambda_{\mathrm{m}}\,\lambda_{\mathrm{n}}\rangle} \frac{\langle 1\,\lambda_{\mathrm{n}}\rangle\,\langle 2\,\lambda_{\mathrm{n}}\rangle}{\langle 1\,\lambda_{\mathrm{m}}\rangle\,\langle 2\,\lambda_{\mathrm{m}}\rangle}.$$

with the $[\![\ \ ]\!]$ brackets all evaluated at $y \in M$. $\mathcal{H}_i^i$ as usual is the reduced determinant obtained after removing the row and column corresponding to the ith positive helicity graviton. Due to the identities

$$\begin{aligned} &\sum_{j=3}^{n} \mathcal{H}_{ij}\,\langle 1\,j\rangle\,\langle 2\,j\rangle + \sum_{\mathrm{m}=1}^{t} \mathcal{H}_{i\mathrm{m}}\,\langle 1\,\lambda_{\mathrm{m}}\rangle\,\langle 2\,\lambda_{\mathrm{m}}\rangle = 0, \\ &\sum_{j=3}^{n} \mathcal{H}_{\mathrm{n}j}\,\langle 1\,j\rangle\,\langle 2\,j\rangle + \sum_{\mathrm{m}=1}^{t} \mathcal{H}_{\mathrm{nm}}\,\langle 1\,\lambda_{\mathrm{m}}\rangle\,\langle 2\,\lambda_{\mathrm{m}}\rangle = 0, \end{aligned} \tag{8.63}$$

and properties of determinants, one can again verify that the choice of positive helicity graviton $i \in \{3, \ldots, n\}$ that we singled out is completely arbitrary.

Collecting all the pieces gives an all-multiplicity formula for the MHV amplitude:

$$\begin{aligned} A_n^{\mathrm{MHV}} = &\sum_{t=0}^{n-4} \sum_{p_1,\ldots,p_t} \int_M \mathrm{vol}_M \frac{\langle 1\,2\rangle^6}{\langle 1\,i\rangle^2\,\langle 2\,i\rangle^2} \left(\prod_{\mathrm{m}=1}^{t} \frac{\mathrm{i}^{-p_{\mathrm{m}}}}{p_{\mathrm{m}}!} \frac{\partial^{p_{\mathrm{m}}}}{\partial \varepsilon_{\mathrm{m}}^{p_{\mathrm{m}}}} \right) |\mathcal{H}_i^i| \Bigg|_{\varepsilon_1=\cdots=\varepsilon_t=0} \\ &\times \exp\left(\mathrm{i} \sum_{j=1}^{n} \mathsf{F}^{\dot{\alpha}}(y, \kappa_j)\,\tilde{\kappa}_{j\,\dot{\alpha}} \right) \prod_{\mathrm{m}=1}^{t} \mathrm{D}\lambda_{\mathrm{m}} \wedge \mathrm{D}\bar{\lambda}_{\mathrm{m}}\, N_{\mathrm{m}}^{(p_{\mathrm{m}}-2)}, \end{aligned} \tag{8.64}$$

having replaced $\mathrm{d}^2 y\,\mathrm{d}^2 \tilde{y}$ in the generating functional (8.36) by a generic volume form vol_M on M (since $\sqrt{|g|} = 1$ in complex coordinates by virtue of the heavenly

Eq. (3.13)). Remembering (8.9), we have also reverted to 'generic' coordinates $y^{\alpha\dot{\alpha}}$ from the complex coordinates $(y^{\dot{\alpha}}, \tilde{y}^{\dot{\alpha}})$, allowing us to uniformly incorporate the plane wave exponentials $\mathrm{e}^{\mathrm{i}[y\,1]}$ and $\mathrm{e}^{\mathrm{i}[\tilde{y}\,2]}$ into the exponential factor on the second line. Lastly, the factor of $\langle 1\,2\rangle^6$ has been reinstated using little group scaling (this scaling property survives turning on curvature in both our momentum eigenstate wavefunctions and our amplitudes).

In this formula, the structure of the SD radiative background space-time M appears through: the diffeomorphism-invariant integration measure; the function $\mathsf{F}^{\dot{\alpha}}(y, \lambda)$ describing the holomorphic curves in the associated twistor space—which appears as the argument of the exponential factor; the dressed momenta in the entries of $\mathcal{H}$; and finally the insertions of (derivatives of) the background news function N arising from tail contributions to the amplitude. The absence of any overall momentum conserving delta functions is an expected feature of scattering in a curved space-time, as the background gravitational field breaks Poincaré invariance. However, a striking feature of the formula is that there is only a *single* residual space-time integral, regardless of the number of external positive helicity gravitons. Perturbatively, general relativity behaves as a cubic field theory (higher-point contact interactions can be re-absorbed into exchange diagrams built from cubic interactions, c.f. [7]), so the naïve expectation is that an n-point tree amplitude in a curved background should entail $n-2$ space-time integrals.

While this simplicity is remarkable from the perspective of perturbation theory based upon the Einstein-Hilbert or Plebanski actions, it is an expected feature of an *MHV formalism* [8] for general relativity, where all vertices are given by MHV interactions, linked by scalar propagators. While the most basic definition of such a formalism (based on all-line shifts) fails [9, 10], there are several indications that an MHV formalism for gravity should exist. At least formally, MHV rules for GR can be defined indirectly by first identifying them in conformal gravity then restricting to Einstein degrees of freedom [11], a truncation which is consistent at tree level [12]. More recently, an off-shell description of general relativity in terms of an action functional on twistor space was found which has a structure compatible with an MHV vertex expansion [13]. The formula (8.64) can be viewed as another smoking gun for gravitational MHV rules, and using the twistor action it should be possible to give a them a precise formulation.

Evaluation on SDPWs: The MHV amplitude (8.64) simplifies considerably when the SD radiative background is a self-dual plane wave (SDPW), where the metric (7.14) is controlled by a single function of lightfront time, $f(y^-)$. The background vertex operators are given on such an SDPW by

$$U^{(p)} = \frac{1}{2\pi\mathrm{i}\hbar}\,\frac{1}{p!}\,\frac{f^{(p-2)}(y^-)\,[m(\iota)\,\tilde{\iota}]^p}{\langle\iota\,1\rangle^2\,\langle\iota\,2\rangle^2}\,, \tag{8.65}$$

where $f^{(k)} \equiv \partial_-^k f(y^-)$ and $y^- := y^{\alpha\dot{\alpha}}\iota_\alpha\tilde{\iota}_{\dot{\alpha}}$. Here, the $\mathbb{P}^1$ integral in the background vertex operator on a general background (8.46) has been performed against the holomorphic delta function in h which localises $\lambda = \iota$. One immediate consequence of the simplicity of (8.65) is that Wick contractions between background vertex operators in an SDPW vanish, dramatically simplifying the structure of the tail contributions. Each insertion of a background vertex operator in the correlator (8.47) must be saturated by contractions with *external* graviton vertex operators. This places a tighter bound on the number of tail contributions to the amplitude for fixed n, with $t \leq \frac{n-3}{2}$.

Using the expression for the twistor curves of an SDPW metric (7.56), it is now straightforward to evaluate (8.64) on this particular class of background. All but one of the four position space integrals can now be done analytically to obtain momentum conserving delta functions in the y^+, z and $\tilde{z}$-directions, leaving

$$A_n^{\text{MHV}} = \delta^3_{+,\perp}\left(\sum_{j=1}^{n} k_j\right) \frac{\langle 1\,2\rangle^6}{\langle 1\,i\rangle^2\,\langle 2\,i\rangle^2} \sum_{t=0}^{\lfloor\frac{n-3}{2}\rfloor} \sum_{p_1,\dots,p_t} \int_{-\infty}^{+\infty} \mathrm{d}y^- \left(\prod_{\text{m}=1}^{t} \frac{\mathrm{i}^{-p_\text{m}}}{p_\text{m}!} \frac{\partial^{p_\text{m}}}{\partial\varepsilon_\text{m}^{p_\text{m}}}\right) |\mathcal{H}^i_i|\Big|_{\varepsilon=0}$$
$$\times\, \mathrm{e}^{\mathrm{i}F_n(y^-)} \prod_{\text{m}=1}^{t} f^{(p_\text{m}-2)}(y^-)\,. \tag{8.66}$$

The three momentum conserving delta functions allow the universal exponent appearing in the second line to be written in the form of a gravitational *Volkov exponent* [14, 15]

$$F_n(y^-) := \int^{y^-} \mathrm{d}s\; g^{ab}(s)\, \frac{\mathbb{K}_a(s)\,\mathbb{K}_b(s)}{2\,\langle\iota|\mathbb{K}|\tilde{\iota}]}\,, \tag{8.67}$$

for $\mathbb{K}_{\alpha\dot{\alpha}}(y^-)$ given by the sum of any (distinct) $n-1$ of the n dressed graviton momenta; for instance

$$\mathbb{K}^{\alpha\dot{\alpha}} = \sum_{i=1}^{n-1} \kappa_i^\alpha\, \tilde{K}_i^{\dot{\alpha}}\,.$$

The block structure (8.61) of the matrix $\mathcal{H}$ remains the same, but with the individual entries simplified:

$$\mathbb{H}_{ij} = \frac{[\![i\,j]\!]}{\langle i\,j\rangle}\,, \quad i \neq j \tag{8.68}$$

$$\mathbb{H}_{ii} = -\sum_{j\neq i} \frac{[\![i\,j]\!]}{\langle i\,j\rangle} \frac{\langle 1\,j\rangle\,\langle 2\,j\rangle}{\langle 1\,i\rangle\,\langle 2\,i\rangle} - \sum_{\text{m}=1}^{t} \varepsilon_\text{m} \frac{[i\,\tilde{\iota}]}{\langle i\,\iota\rangle} \frac{\langle 1\,\iota\rangle\,\langle 2\,\iota\rangle}{\langle 1\,i\rangle\,\langle 2\,i\rangle}\,,$$

$$\mathfrak{h}_{i\text{m}} = \varepsilon_\text{m} \frac{[i\,\tilde{\iota}]}{\langle i\,\iota\rangle}\,, \qquad \mathbb{T}_{\text{mn}} = -\delta_{\text{mn}}\,\varepsilon_\text{m} \sum_{i=1}^{n} \frac{[\tilde{\iota}\,i]}{\langle\iota\,i\rangle} \frac{\langle 1\,i\rangle\,\langle 2\,i\rangle}{\langle 1\,\iota\rangle\,\langle 2\,\iota\rangle}\,.$$

The $[\![i\,j]\!] = \tilde{K}_i^{\dot{\alpha}} \tilde{K}_{j\dot{\alpha}}$ are now simply contractions of the dressed spinors $\tilde{K}_i^{\dot{\alpha}}$ defined as in (7.60). The formula (8.66) is manifestly diffeomorphism invariant, and is separately symmetric in the positive and the negative helicity gravitons.

We can easily write down some low point examples that bring out the structure of our formulae. The 3-graviton MHV amplitude is found to be

$$A(1^- 2^- 3^+) = \delta^3_{+,\perp}\left(\sum_{j=1}^{3} k_j\right) \frac{\langle 1\,2\rangle^6}{\langle 2\,3\rangle^2\,\langle 3\,1\rangle^2} \int \mathrm{d}y^-\ \mathrm{e}^{\mathrm{i}F_3}\,, \tag{8.69}$$

which is almost identical to what one finds in flat space. At 4 points, background dressed momenta enter the game,

$$A(1^- 2^- 3^+ 4^+) = \delta^3_{+,\perp}\left(\sum_{j=1}^{4} k_j\right) \int \mathrm{d}y^-\ \mathrm{e}^{\mathrm{i}F_4}\ \frac{\langle 1\,2\rangle^6\,[\![3\,4]\!](y^-)}{\langle 1\,3\rangle\,\langle 1\,4\rangle\,\langle 2\,3\rangle\,\langle 2\,4\rangle\,\langle 3\,4\rangle}\,. \tag{8.70}$$

At 5 points and higher, one starts observing tail terms and explicit expressions become more involved. Nonetheless, it is possible to compute all the ε_{m} derivatives in (8.66). This results in a cumbersome but explicit formula and was described in [14] and in Appendix B of [16].

8.3 Consistency Checks

While the unitarity methods which can be used to prove tree-level amplitude formulae in vacuum no longer apply in the presence of a strong background, there are still some basic consistency checks which can be run on formula (8.64). The first of these is comparison with explicit Feynman diagram calculations using the background-coupled Einstein-Hilbert Lagrangian; of course, such computations are only tractable at low numbers of points and in highly symmetric SD radiative space-times where the Feynman rules in the background can be determined explicitly. Nevertheless, in Appendix A of [16], we have explicitly demonstrated that for an SDPW background, the formula (8.66) matches the amplitude computed with Feynman rules at 3- and 4-points.

Beyond this, the most straightforward all-multiplicity consistency check is the flat space limit, where the SD radiative background becomes (complexified) Minkowski space. In this case, the background news function $N(u,\lambda,\bar{\lambda})$ becomes trivial so only the $t=0$, no-tail term in (8.64) contributes. For Minkowski space, the twistor curves become twistor lines:

$$\mathsf{F}^{\dot\alpha}(y,\lambda)|_{\text{flat}} = y^{\alpha\dot\alpha}\,\lambda_\alpha\,. \tag{8.71}$$

This enables all of the remaining integrals in (8.64) to be performed immediately, with the result

$$\delta^4\left(\sum_{i=1}^{n} k_i\right) \frac{\langle 1\,2\rangle^6}{\langle 1\,i\rangle^2\,\langle 2\,i\rangle^2}\,\left|\mathbb{H}^i_i\right|\,, \tag{8.72}$$

where the entries of $\mathbb{H}$ are now

$$\mathbb{H}_{ij} = \frac{[i\,j]}{\langle i\,j\rangle}\,, \quad i \neq j\,, \qquad \mathbb{H}_{ii} = -\sum_{j\neq i} \frac{[i\,j]}{\langle i\,j\rangle}\,\frac{\langle 1\,j\rangle\,\langle 2\,j\rangle}{\langle 1\,i\rangle\,\langle 2\,i\rangle}\,, \tag{8.73}$$

defined in terms of un-dressed on-shell momenta. The expression (8.72) is precisely Hodges' formula (5.41) for graviton scattering in flat space, so we do indeed produce the correct flat space limit.

A more non-trivial consistency check is the behaviour of (8.64) in the *perturbative limit*, where the SD radiative background becomes weak and is treated like a single positive helicity graviton in flat space. In the perturbative limit, any SD radiative space-time will be well-approximated by a SDPW, so it suffices to work directly with (8.66) and isolate all contributions to the MHV amplitude which are linear in the background profile $f(y^-)$. This background-dependence enters (8.66) in three ways: through the matrix $\mathcal{H}$, whose entries include dressed momentum spinors; through the Volkov exponent (8.67); or through explicit tail factors when $t > 0$. Clearly, only the $t = 0, 1$ terms can provide such linear contributions.

In the $t = 0$ term, linear dependence on $f(y^-)$ arises only from the Volkov exponent or dressed momentum spinors. This is easily determined explicitly using (7.60), and representing the perturbative background by the Fourier mode

$$f(y^-) = \kappa\,\mathrm{e}^{-\mathrm{i}\,\omega\,y^-} := \kappa\,\mathrm{e}^{\mathrm{i}\,q\cdot y}\,, \qquad q_{\alpha\dot\alpha} = \omega\,\iota_\alpha\,\tilde\iota_{\dot\alpha}\,, \tag{8.74}$$

the contributions to the perturbative limit of the $t = 0$ term are

$$\frac{\langle 1\,2\rangle^6}{\langle 1\,i\rangle^2\,\langle 2\,i\rangle^2}\,\delta^4\!\left(q + \sum_{i=1}^{n} k_i\right)\left[\sum_{\substack{j,l=3\\ j\neq l,\ j,l\neq i}}^{n} (-1)^{j+l}\,\frac{[\tilde\iota\,j]\,[\tilde\iota\,l]}{\langle \iota\,j\rangle\,\langle \iota\,l\rangle}\,|\mathbb{H}^{ij}_{il}| - |\mathbb{H}^{i}_{i}|\sum_{j=1}^{n}\frac{[\tilde\iota\,j]\,\langle o\,j\rangle^2}{\langle \iota\,j\rangle} \right.$$
$$\left. - \sum_{\substack{j=3\\ j\neq i}}^{n} |\mathbb{H}^{ij}_{ij}| \sum_{l\neq j} \frac{[\tilde\iota\,j]\,[\tilde\iota\,l]}{\langle \iota\,j\rangle\,\langle \iota\,l\rangle}\,\frac{\langle 1\,l\rangle\,\langle 2\,l\rangle}{\langle 1\,j\rangle\,\langle 2\,j\rangle}\right], \tag{8.75}$$

where $\mathbb{H}$ is now understood to be the 'flat' matrix given by (8.73), and $\mathbb{H}^{i\cdots j}_{k\cdots l}$ denotes the submatrix obtained by removing rows $i, \ldots, j$ and columns $k, \ldots, l$. Using the Schouten identity and 4-momentum conservation, these terms can be re-written as

$$\frac{\langle 1\,2\rangle^6}{\langle 1\,i\rangle^2\,\langle 2\,i\rangle^2}\,\delta^4\!\left(q+\sum_{i=1}^{n}k_i\right)\left[-|\mathbb{H}^i_i|\sum_j\frac{[\tilde{\iota}\,j]}{\langle\iota\,j\rangle}\frac{\langle 1\,j\rangle\,\langle 2\,j\rangle}{\langle 1\,\iota\rangle\,\langle 2\,\iota\rangle}+\sum_{j,l,j\neq l}(-1)^{j+l}\frac{[\tilde{\iota}\,j]\,[\tilde{\iota}\,l]}{\langle\iota\,j\rangle\,\langle\iota\,l\rangle}\,|\mathbb{H}^{ij}_{il}|\right.$$
$$\left.-\sum_j|\mathbb{H}^{ij}_{ij}|\sum_{l\neq j}\frac{[\tilde{\iota}\,j]\,[\tilde{\iota}\,l]}{\langle\iota\,j\rangle\,\langle\iota\,l\rangle}\,\frac{\langle 1\,l\rangle\,\langle 2\,l\rangle}{\langle 1\,j\rangle\,\langle 2\,j\rangle}\right],\tag{8.76}$$

where the precise ranges of the various summations are implied by the structure of the summands.

The $t=1$ term only contributes to the perturbative limit through the explicit tail insertion itself; all other background-dependence (in the matrix $\mathcal{H}$ and Volkov exponent) is set to zero when extracting those terms linear in $f(y^-)$. The resulting contribution is

$$\frac{\langle 1\,2\rangle^6}{\langle 1\,i\rangle^2\,\langle 2\,i\rangle^2}\sum_{p=1}^{n-3}\sum_{j_1,\ldots,j_p}\left[\sum_{k,l}(-1)^{k+l}\left|\mathbb{H}^{ikj_1\cdots j_p}_{ilj_1\cdots j_p}\right|\left(\prod_{\mathrm{m}=1}^{p}\frac{[\tilde{\iota}\,j_{\mathrm{m}}]}{\langle\iota\,j_{\mathrm{m}}\rangle}\frac{\langle 1\,\iota\rangle\,\langle 2\,\iota\rangle}{\langle 1\,j_{\mathrm{m}}\rangle\,\langle 2\,j_{\mathrm{m}}\rangle}\right)\frac{[\tilde{\iota}\,k]\,[\tilde{\iota}\,l]}{\langle\iota\,k\rangle\,\langle\iota\,l\rangle}\right.$$
$$\left.+\left|\mathbb{H}^{ij_1\cdots j_p}_{ij_1\cdots j_p}\right|\left(\prod_{\mathrm{m}=1}^{p}\frac{[\tilde{\iota}\,j_{\mathrm{m}}]}{\langle\iota\,j_{\mathrm{m}}\rangle}\frac{\langle 1\,\iota\rangle\,\langle 2\,\iota\rangle}{\langle 1\,j_{\mathrm{m}}\rangle\,\langle 2\,j_{\mathrm{m}}\rangle}\right)\sum_l\frac{[\tilde{\iota}\,l]}{\langle\iota\,l\rangle}\frac{\langle 1\,l\rangle\,\langle 2\,l\rangle}{\langle 1\,\iota\rangle\,\langle 2\,\iota\rangle}\right],\tag{8.77}$$

where the momentum conserving delta functions (same as those appearing in (8.76)) have been omitted and $\mathbb{H}$ is once again understood as the matrix with 'flat' entries.

The $t=0$ and $t=1$ contributions combine auspiciously if we define an $(n-1)\times(n-1)$ matrix $\widehat{\mathbb{H}}$ whose entries are equivalent to those of $\mathbb{H}$, with the extra row and column corresponding to the background graviton with momentum $q_{\alpha\dot{\alpha}}$. Now, the minor $|\widehat{\mathbb{H}}^q_q|$ differs from $|\mathbb{H}|$ only by the diagonal entries of the two matrices, so

$$|\widehat{\mathbb{H}}^q_q|=|\mathbb{H}|+\sum_{p=1}^{n-2}\sum_{j_1,\ldots,j_p}\left|\mathbb{H}^{j_1\cdots j_p}_{j_1\cdots j_p}\right|\left(\prod_{\mathrm{m}=1}^{p}\frac{[\tilde{\iota}\,j_{\mathrm{m}}]}{\langle\iota\,j_{\mathrm{m}}\rangle}\frac{\langle 1\,\iota\rangle\,\langle 2\,\iota\rangle}{\langle 1\,j_{\mathrm{m}}\rangle\,\langle 2\,j_{\mathrm{m}}\rangle}\right).\tag{8.78}$$

Applying this identity to further reduced minors of $\mathbb{H}$, it follows that the $t=0,1$ contributions to the perturbative limit combine to give

$$\delta^4\!\left(q+\sum_{i=1}^{n}k_i\right)\frac{\langle 1\,2\rangle^6}{\langle 1\,i\rangle^2\,\langle 2\,i\rangle^2}\,|\widehat{\mathbb{H}}^i_i|\,,\tag{8.79}$$

which is the Hodges formula for MHV graviton scattering in flat space with $(n-1)$ positive helicity gravitons whose momenta are $(k_3,\ldots,k_n,q)$. This confirms that (8.64) has the correct perturbative limit, emerging only after a somewhat subtle collaboration between 'pure scattering' (i.e., $t=0$) and tail (i.e., $t=1$) contributions to the amplitude.

References

1. Mason LJ, Skinner D (2010) Gravity, twistors and the MHV formalism. Commun Math Phys 294:827–862 arXiv:0808.3907
2. Friedlander FG (1975) The wave equation on a curved space-time. Cambridge University Press
3. Harte AI (2013) Tails of plane wave spacetimes: wave-wave scattering in general relativity. Phys Rev D88(8):084059. arXiv:1309.5020
4. Stanley RP (1999) Enumerative combinatorics, vol 2. Cambridge University Press
5. van Lint JH, Wilson RP (2001) A course in combinatorics, vol 2. Cambridge University Press
6. Stanley RP (2012) Topics in algebraic combinatorics. http://www-math.mit.edu/~rstan/algcomb.pdf
7. Bern Z, Carrasco JJ, Chiodaroli M, Johansson H, Roiban R (2019) The duality between color and kinematics and its applications. arXiv:1909.01358
8. Cachazo F, Svrcek P, Witten E (2004) MHV vertices and tree amplitudes in gauge theory. JHEP 09:006 hep-th/0403047
9. Bjerrum-Bohr NEJ, Dunbar DC, Ita H, Perkins WB, Risager K (2006) MHV-vertices for gravity amplitudes. JHEP 01:009 hep-th/0509016
10. Bianchi M, Elvang H, Freedman DZ (2008) Generating Tree Amplitudes in N = 4 SYM and N = 8 SG. JHEP 09:063 arXiv:0805.0757
11. Adamo T, Mason L (2014) Conformal and Einstein gravity from twistor actions. Class Quant Grav 31(4):045014. arXiv:1307.5043
12. Maldacena J (2011) Einstein gravity from conformal gravity. arXiv:1105.5632
13. Sharma A (2021) Twistor action for general relativity. arXiv:2104.07031
14. Adamo T, Mason L, Sharma A (2020) MHV scattering of gluons and gravitons in chiral strong fields. Phys Rev Lett 125(4):041602. arXiv:2003.13501
15. Adamo T, Ilderton A (2020) Classical and quantum double copy of back-reaction. JHEP 09:200 arXiv:2005.05807
16. Adamo T, Mason L, Sharma A (2022) Graviton scattering in self-dual radiative space-times. arXiv:2203.02238

Chapter 9
Outlook

Over the course of this thesis, twistor sigma models have provided us with many new insights into classical and perturbative gravity. Through Theorem 3.2, they provided a concrete formula to translate hamiltonian complex structure deformations of twistor space into hyperkähler metrics on SD vacuum space-times. Over the course of Chaps. 5–8, we saw the applications of Theorem 3.2 to the computation of MHV graviton amplitudes as well as their holographic symmetries. Using semi-classical correlators of our models, we derived the well-established Hodges' formula (5.41) in flat space. Since our proof directly linked space-time perturbation theory to sigma model correlators, we were able to extend this calculation to interesting classes of self-dual space-times. This resulted in the formula (8.64) for MHV graviton scattering on SD radiative space-times, which we made much more explicit in formula (8.66) for our running example of SD plane wave backgrounds.

Amplitudes in curved backgrounds. The SD radiative backgrounds we dealt with are toy models for backgrounds encountered in the real world. But nevertheless, they allow us to see past the maximal symmetries of flat space and extract universal mathematical structures governing perturbative QFT. There may be next to no global symmetries in these backgrounds, but the magic of twistor theory still prevails and generates miraculously simple expressions for amplitudes. Similar expressions for N^kMHV amplitudes can be conjectured (though not yet proven) using generalizations of our sigma models that govern higher degree rational curves [1]. Combined with similar formulae for gluon scattering derived in [2], our results provide the first steps toward developing on-shell methods for perturbative computations in the presence of strong background fields.

While the structure of graviton scattering in a curved SD radiative space-time has many interesting features, the structure of tail contributions to the amplitude (encoded by the background news function) seems particularly noteworthy. Analogous 'tail effects' also arise in the classical limit of gravitational scattering [3–7], and have been the subject of study recently in the context of black hole scattering (e.g., [8–10]). It

A. Sharma, *Twistor Sigma Models*, Springer Theses,
https://doi.org/10.1007/978-3-031-50751-9_9

would be interesting to establish what—if any—relationship there is between these notions of gravitational 'tails', and if our all-multiplicity results can be of any use in the context of early-inspiral black hole physics.

Our results provide a variety of avenues for future work. They act as a proof of concept that twistor theory underlies the simplicity in scattering amplitudes even when other methods to compute amplitudes may not apply. More optimistically, they give us hope that other tools of the on-shell amplitudologist like BCFW recursion, generalized unitarity, etc. might also extend to curved backgrounds. Self-dual backgrounds provide a perfect setting to test how far these tools stretch, as our twistor sigma models already provide us with at least one means of computing the tree amplitudes exactly (at least in the MHV case). In more practical directions, the results of our work pave the way for efficient computations of amplitudes in more commonly encountered strong backgrounds like cosmology, AdS/CFT, strong field QED/QCD, black holes, etc. For example, it is already possible to systematically treat scattering on Gibbons-Hawking instanton space-times by treating them as radiative space-times in $(2, 2)$ signature [1, 11], and in principle our calculations could also be extended to compute amplitudes in self-dual black hole backgrounds.

Moving beyond self-duality brings its own set of hurdles. Generic non-self-dual backgrounds are not hyperkähler, so our twistor sigma models can no longer capture their dynamics. Nonetheless, in recent years the tools to tackle such backgrounds have been provided by more general string theories [12] in *ambitwistor space*: the space of null geodesics in space-time. At tree-level, these ambitwistor strings can be consistently coupled to general on-shell backgrounds in supergravity theories [13, 14]. Worldsheet correlators of these models then compute amplitudes around such backgrounds. The feasibility of using ambitwistor strings to compute amplitudes has been demonstrated on plane wave backgrounds at three points [15], and somewhat formally in (A)dS backgrounds at general multiplicity for certain theories [16–19]. It would be interesting to pursue this approach to build scattering equations and worldsheet formulae for amplitudes on other space-times of interest.

Flat space holography. The study of celestial holography is still in a nascent stage. Apart from their practical applications to the study of scattering amplitudes, our twistor sigma models also played an important role in connecting celestial holography to twistor theory. Using them, we proved in Chap. 6 that the loop algebra of $w_{1+\infty}$ observed in the positive helicity subsector of soft graviton collinear limits originated in the symmetries of self-dual GR. These then translated to symmetries of the OPE algebra of celestial CFT. Since celestial holography is constrained by such an infinite dimensional symmetry algebra, it is natural to wonder if the symmetries can be used to bootstrap its CFT data even in the absence of concrete realizations of CCFT.

In recent years, the search for explicit constructions of CCFT through string theoretic means has also begun to yield interesting insights. Preliminary investigations [20] based on holography applied to topological strings on twistor space display signs of deep connections between twistor theory and stringy realizations of celestial CFTs. For example, if one studies the type I topological B model [21] with twistor space $\mathbb{PT} = \mathscr{O}(1) \oplus \mathscr{O}(1)$ as target space, one can obtain toy models of celestial holography [22, 23] via D1 brane back-reaction effects much as when twisted holography is applied to the resolved conifold $\mathcal{O}(-1) \oplus \mathcal{O}(-1)$ [24]. The resulting dual (defect) CFTs are given by chiral symplectic boson systems of the same nature as our sigma models. On the back-reacted space-time, this example gives rise to a theory of gravity that is not GR but instead governs self-dual Kähler (but not necessarily Ricci-flat) metrics. It is hoped that our twistor sigma models will act as building blocks for future generalizations of twisted holography to self-dual GR (and beyond).

Towards quantization: The twistor sigma model (3.49) gives rise to gravitational amplitudes via its classical action and the corresponding tree expansion; by contrast twistor strings or ambitwistor strings produce amplitudes as fully quantum correlations functions in the worldsheet CFT. This distinction leaves room for one to ask what the twistor sigma model could correspond to if treated quantum mechanically. In particular, there is scope for this to give rise to some theory of self-dual quantum gravity, for instance as envisaged by [25, 26] for the $\mathcal{N} = 2$ string. For instance, the 'quantum' (i.e., at finite $\hbar$ as well as containing contributions from disconnected diagrams) MHV graviton amplitude produced by the twistor sigma model can be computed [27]:

$$\delta^4\left(\sum_{r=1}^{n} k_r\right) \langle 1\,2\rangle^{2n} \prod_{i=3}^{n} \frac{1}{\langle 1\,i\rangle^2 \langle 2\,i\rangle^2} \exp\left[-\frac{i\hbar}{8\pi}\sum_{j\neq i} \frac{[i\,j]}{\langle i\,j\rangle} \frac{\langle 1\,i\rangle^2 \langle 2\,j\rangle^2}{\langle 1\,2\rangle^2}\right], \tag{9.1}$$

although its physical properties and interpretation remain to be explored. It would also be intriguing to make contact with the $*$-algebra definition of the quantum $W_{1+\infty}$-algebra as described in [28] and the Moyal deformations of the Poisson structure associated to self-dual gravity proposed by [29] which are closely connected also to Penrose's Palatial twistor ideas [30]. It would be interesting to track the twistor-theoretic component of the other appearances of W-infinity algebras in the literature.

References

1. Adamo T, Mason L, Sharma A (2022) Graviton scattering in self-dual radiative space-times. arXiv:2203.02238
2. Adamo T, Mason L, Sharma A (2020) Gluon scattering on self-dual radiative gauge fields. arXiv:2010.14996
3. Bonnor WB (1959) Spherical gravitational waves. Phil Trans Roy Soc Lond A A 251:233–271
4. Bonnor WB, Rotenberg MA (1966) Gravitational waves from isolated sources. Proc Roy Soc Lond A A 289:247–274
5. Thorne KS (1980) Multipole expansions of gravitational radiation. Rev Mod Phys 52:299–339
6. Blanchet L, Damour T (1988) Tail transported temporal correlations in the dynamics of a gravitating system. Phys Rev D 37:1410
7. Blanchet L, Schaefer G (1993) Gravitational wave tails and binary star systems. Class Quantum Gravity 10:2699–2721
8. Foffa S, Sturani R (2013) Tail terms in gravitational radiation reaction via effective field theory. Phys Rev D 87(4):044056. arXiv:1111.5488

9. Galley CR, Leibovich AK, Porto RA, Ross A (2016) Tail effect in gravitational radiation reaction: time nonlocality and renormalization group evolution. Phys Rev D 93:124010 [arXiv:1511.07379]
10. Bern Z, Parra-Martinez J, Roiban R, Ruf MS, Shen C-H, Solon MP, Zeng M (2021) Scattering amplitudes, the tail effect, and conservative binary dynamics at $O(G^4)$. arXiv:2112.10750
11. Tod KP, Ward RS (1979) Selfdual metrics with selfdual Killing vectors. Proc Roy Soc Lond A 368:411–427
12. Mason L, Skinner D (2014) Ambitwistor strings and the scattering equations. JHEP 07:048 [arXiv:1311.2564]
13. Adamo T, Casali E, Skinner D (2015) A worldsheet theory for supergravity. JHEP 02:116 [arXiv:1409.5656]
14. Adamo T, Casali E, Nekovar S (2019) Ambitwistor string vertex operators on curved backgrounds. JHEP 01:213 [arXiv:1809.04489]
15. Adamo T, Casali E, Mason L, Nekovar S (2017) Amplitudes on plane waves from ambitwistor strings. JHEP 11:160 [arXiv:1708.09249]
16. Roehrig K, Skinner D (2022) Ambitwistor strings and the scattering equations on $AdS_3 \times S^3$. JHEP 02:073 [arXiv:2007.07234]
17. Eberhardt L, Komatsu S, Mizera S (2020) Scattering equations in AdS: scalar correlators in arbitrary dimensions. JHEP 11:158 [arXiv:2007.06574]
18. Gomez H, Jusinskas RL, Lipstein A (2021) Cosmological scattering equations. Phys Rev Lett 127(25):251604. arXiv:2106.11903
19. Gomez H, Jusinskas RL, Lipstein A (2021) Cosmological scattering equations at tree-level and one-loop. arXiv:2112.12695
20. Costello K, Paquette NM (2022) Celestial holography meets twisted holography: 4d amplitudes from chiral correlators. arXiv:2201.02595
21. Costello K, Li S (2020) Anomaly cancellation in the topological string. Adv Theor Math Phys 24(7):1723–1771. arXiv:1905.09269
22. Costello K, Paquette NM, Sharma A (2023) Top-down holography in an asymptotically flat spacetime. Phys Rev Lett 130(6):061602. arXiv:2208.14233
23. Costello K, Paquette NM, Sharma A (2023) Burns space and holography. arXiv:2306.00940
24. Costello K, Gaiotto D (2018) Twisted holography. arXiv:1812.09257
25. Ooguri H, Vafa C (1990) Selfduality and $N = 2$ string magic. Mod Phys Lett A 5:1389–1398
26. Ooguri H, Vafa C (1991) Geometry of $N = 2$ strings. Nucl Phys B 361:469–518
27. Adamo T, Mason L, Sharma A (2021) Twistor sigma models for quaternionic geometry and graviton scattering. arXiv:2103.16984
28. Pope CN (1991) Lectures on W algebras and W gravity. In: Summer school in high-energy physics and cosmology, vol 12, pp. 827–867. hep-th/9112076
29. Strachan IAB (1992) The Moyal algebra and integrable deformations of the selfdual Einstein equations. Phys Lett B 283:63–66
30. Penrose R (2015) Palatial twistor theory and the twistor googly problem. Phil Trans Roy Soc Lond A 373:20140237

SPRINGER NATURE

GPSR Compliance

The European Union's (EU) General Product Safety Regulation (GPSR) is a set of rules that requires consumer products to be safe and our obligations to ensure this.

If you have any concerns about our products, you can contact us on ProductSafety@springernature.com

In case Publisher is established outside the EU, the EU authorized representative is:

Springer Nature Customer Service Center GmbH
Europaplatz 3
69115 Heidelberg, Germany

The manufacturer's authorised representative in the EU is Springer Nature Customer Service Centre GmbH, Europaplatz 3, 69115 Heidelberg, Germany. If you have any concerns regarding our products, please contact ProductSafety@springernature.com

Printed and bound by CPI Group (UK) Ltd, Croydon, CR0 4YY

01/10/2026

02225876-0006